Foundations in a Fertile Soil

Farming and Farm Buildings in Lancaster County, Pennsylvania

The rich panorama of Lancaster County's farms and farm buildings is captured in this photograph by Ed Worteck

Foundations in a Fertile Soil

Farming and Farm Buildings in Lancaster County, Pennsylvania

David B. Schneider

The Historic Preservation Trust of Lancaster County

Principal Photography by **Ed Worteck**

Essay *The Pennsylvania Barn* by **Robert F. Ensminger**

Historic Preservation Trust
of Lancaster County

For my son Michael and his generation to whom this legacy will be passed and without whom there would be no need for historic preservation.

The Historic Preservation Trust of Lancaster County
123 N. Prince Street
Lancaster, Pennsylvania, 17603

Photograph credits:
Principal photography by Ed Worteck.
Additional photographs by John P. Herr, Robert Lowing, Heidi M. Pawlowski, David B. Schneider, and Lori Stahl.

Cover Photo by Lori Stahl.

ISBN 0-9635153-1-4
Printed in the United States of America
by The Village Press of Bird-in-Hand, Inc.

First Edition

Contents

Acknowledgments

This project was funded in part with Federal funds from the National Park Service, Department of the Interior. These funds were administered as a grant through the Pennsylvania Historical and Museum Commission (Bureau for Historic Preservation). Thanks are extended to the staff of the Bureau for their help and guidance, plus their patience, especially to Greg Ramsey, Jerry Clouse, and Janice Stramara.

We extend our sincere thanks and appreciation to the owners of the six properties used as case studies in this book, both for their cooperation with our efforts and for their excellent stewardship of these important sites: Mr. Paul Martin, the Habecker family, Mrs. William H. Paul, G. Gerald Beane and Holly Doench, Mrs. Elizabeth Shreiner and her family, Mr. and Mrs. P. Kenneth Rohrer and their family, and Mr. and Mrs. Donald H. Herr. Thanks are also extended to all the property owners of sites illustrated herein or that were studied as part of the development of this publication.

A great many people were involved in the the preparation of this book, which represents the culmination of several major activities of the Historic Preservation Trust of Lancaster County's *Rural Preservation Project*. To all of them we extend our appreciation: Ed Worteck donated his considerable artistic talent and vision, plus a great deal of his time and energy, to producing photographs for this book. Heidi M. Pawlowski served as the Trust's Rural Preservation Specialist in 1992 and 1993 and was responsible for much of the field survey and research work on which this book is based. Ed Shock provided volunteer time preparing additional background research. David Schuyler, Professor of American Studies at Franklin & Marshall College, shared his considerable editorial talent. Tom Daniels, Joseph Glass, Clarke Hess, and Steve Miller, provided extensive input into the various drafts of the text. Carey Shaner and Nancy Haubert edited the final draft. Jim Peterson, of the Art Department at Franklin & Marshall College, assisted with the graphic design of this book.

The Board of Directors of the Historic Preservation Trust of Lancaster County recognized the need for the Rural Preservation Project and gave their staff considerable latitude in accomplishing its agenda.

The members of Historic Preservation Trust's Rural Preservation Project Advisory Committee provided invaluable input, assistance, and advice throughout the project: Gary Baer, Christian Earl Eaby, Joseph Glass, Clarke Hess, James Kurtz, and Alan Musselman.

Additional funding was provided by the following corporate donors:

Armstrong World Industries

The Building Industry Association of Lancaster

The Lancaster County Developers Association

This project has been financed in part with Federal funds from the National Park Service, Department of the Interior. However, the contents and opinions do not necessarily reflect the views or policies of the Department of the Interior, nor does the mention of trade names or commercial products constitute endorsement or recommendation by the Department of Interior.

This program receives Federal funds from the National Park Service. Under Title VI of the Civil Rights Act of 1964 and Section 504 of the Rehabilitation Act of 1973, the U.S. Department of the Interior strictly prohibits unlawful discrimination in departmental Federally Assisted Programs on the basis of race, color, national origin, age or handicap. Any person who believes he or she has been discriminated against should write to: Director, Equal Opportunity Program, Department of the Interior, National Park Service, P. O. Box 37127, Washington, DC 20013-7127.

Introduction

Lancaster County, Pennsylvania, is a historic and cultural treasure of national, if not international, significance. What defines it as such is the combination of rich farmland and the cultural traditions of the people who lived and worked on that land for almost three hundred years. It is made tangible by a rich and varied landscape of farms and farm buildings.

Lancaster County has rightfully been called "the Garden Spot of America." Since 1710, farming has dominated the local culture and economy, first as a subsistence activity and soon a commercial one. By the late 1700s, Lancaster was the top agricultural producing county in Pennsylvania, then referred to as "the bread basket of the colonies" for its extensive wheat production. As wheat growing was yielded to farms further west in the mid-1800s, Lancaster County farmers turned to tobacco and cattle. The county continued its leadership in agricultural production, topping all other counties in the nation in agricultural censuses from 1850 to 1900. It was not until the early 1900s that the county's dominance was challenged and eventually overtaken by irrigated, and much larger, counties in California. Yet, Lancaster County remains a place whose character is largely defined by farming and that continues to lead the nation in many statistical categories of farm output.

Thomas W. Kemp summarized the role of agriculture in the county's history in H. M. J.

The Garden Spot of America

The character of Lancaster County is largely defined by its expanses of fertile farmland, its historic farmsteads, and the cultural influences of its people, including Pennsylvania Dutch, Amish, Mennonites, and others. • *Lori Stahl*

Klein's 1924 *Lancaster County Pennsylvania: A History:*

So closely is agriculture woven into the history of Lancaster county that every phase of development has for its basis the unfolding and strengthening of agricultural evolution. The history of every township, every borough and of Lancaster city is part and parcel of the agricultural history of the county as a whole. The county was founded upon agriculture and to-day it is the determining factor in every progressive step of its history. From the standpoint of agricultural wealth Lancaster county to-day produces more than any single area of equal size in the United States. It is the sustaining force that gives to the industries of the county their peculiar advantage over the industries of other localities.

The cultural influences that shaped Lancaster County and contributed to its leadership in agriculture add to the overall significance of its agricultural history. Pennsylvania Germans, English, French, Welsh, Scotch-Irish, and others contributed to the diversity of the county's character. The county's plain sect population is an essential element of its cultural composition and the historical and continued presence of substantial Old Order Amish, Mennonite, and other plain sects is recognized nationally as the very symbol of the community.

This book is both an end and a beginning. As an end, it culminates four years work as part of the Rural Preservation Project of the Historic Preservation Trust of Lancaster County. This project included surveys of thousands of farming related buildings, extensive research, and the preparation of a National Register of Historic Places Multiple Property Documentation Form, as well nominations for six individual farms, under the title "The Historic Farming Resources of Lancaster County." The text of this book has been edited directly from these National Register forms.

The book is a beginning in that it has only scratched the surface in documenting the immense contribution that farming and farming related buildings have made to the county, state, and nation. One of the problems encountered in the preparation of this study was the lack of a detailed published history of farming in Lancaster County. This is surprising considering the significance of agriculture to the county and its national recognition. The first section of this book represents, to our knowledge, the most detailed outline of the history of Lancaster County farming yet available.

It is our hope that this book and the National Register nominations on which it is based will do more than simply record Lancaster County farming and its architecture. It is easy to take the commonplace for granted. Yet it is the commonplace in Lancaster County, its farms and farm buildings, that are part of the living presence that defines this place as a national treasure. These important community resources are threatened with extinction, and with them will go the rich quality of life Lancastrians enjoy. We hope this effort will help to stimulate increased interest and appreciation for these aspects of our cultural heritage. We believe this will be an important step toward preserving the legacy of farming and farming related buildings that was passed to our generation and that we hold in trust for the future.

David B. Schneider
Executive Director
Historic Preservation Trust of Lancaster County

A Note on the Photographs

In producing this book, we hoped to both record the historic architecture of Lancaster County and to capture the spirit of its sense of place. The latter requires the sensitivity and perception of an artist who understands and truly appreciates the subject matter. Photographer Ed Worteck, who donated most of the professional photographs in this book, is such a person. Mr. Worteck, Chair of the Art Department at Goucher College in Maryland, has been undertaking an extensive project to photograph the cultural and historic character of Lancaster County. The photographs included here represent only a portion of his work.

The talents of noted local photographers John P. Herr, Robert Lowing, and Lori Stahl, are also represented.

1: Farming in Lancaster County

Since before the American Revolution, Lancaster County, Pennsylvania, has been noted for its significant contribution to agriculture, eventually earning the nickname "The Garden Spot of America."[1] As the principal historic economic activity of the county, farming has both influenced and has been influenced by the development of other industries, transportation networks, and the cultural patterns of the county's residents.

Farming has shaped the social, cultural, and political history of the county and defines much of its present character. The county's farms continue to reflect this rich history, which extends from the period of early settlement and subsistence farming, through its prominence in wheat production, its dominance in tobacco production, and its adaptation to changing technologies and market conditions that continues today. Lancaster County's nationally recognized community character is defined by its farming landscape, its farm buildings, and the history, culture, and traditions of the people who created, used, and continue to use them. Among these people are the Pennsylvania Germans (or "Dutch"), including plain sects, principally Old Order Amish and Mennonites, who historically dominated the local farming culture and who are so closely identified with its current sense of place. English Quaker, Scotch-Irish, French, and others added cultural diversity and also significantly influenced Lancaster's character and history.[2]

• *Ed Worteck*

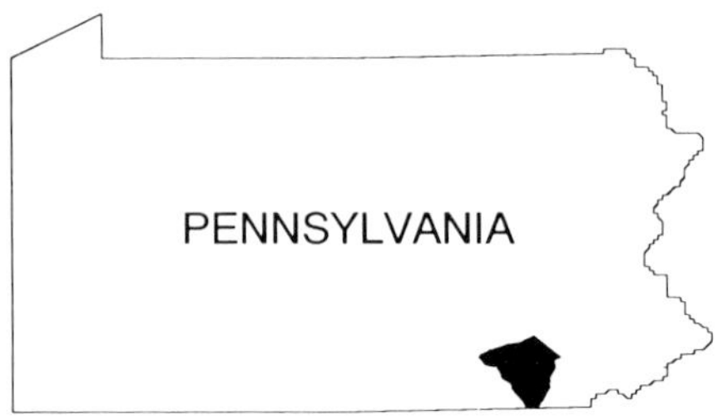

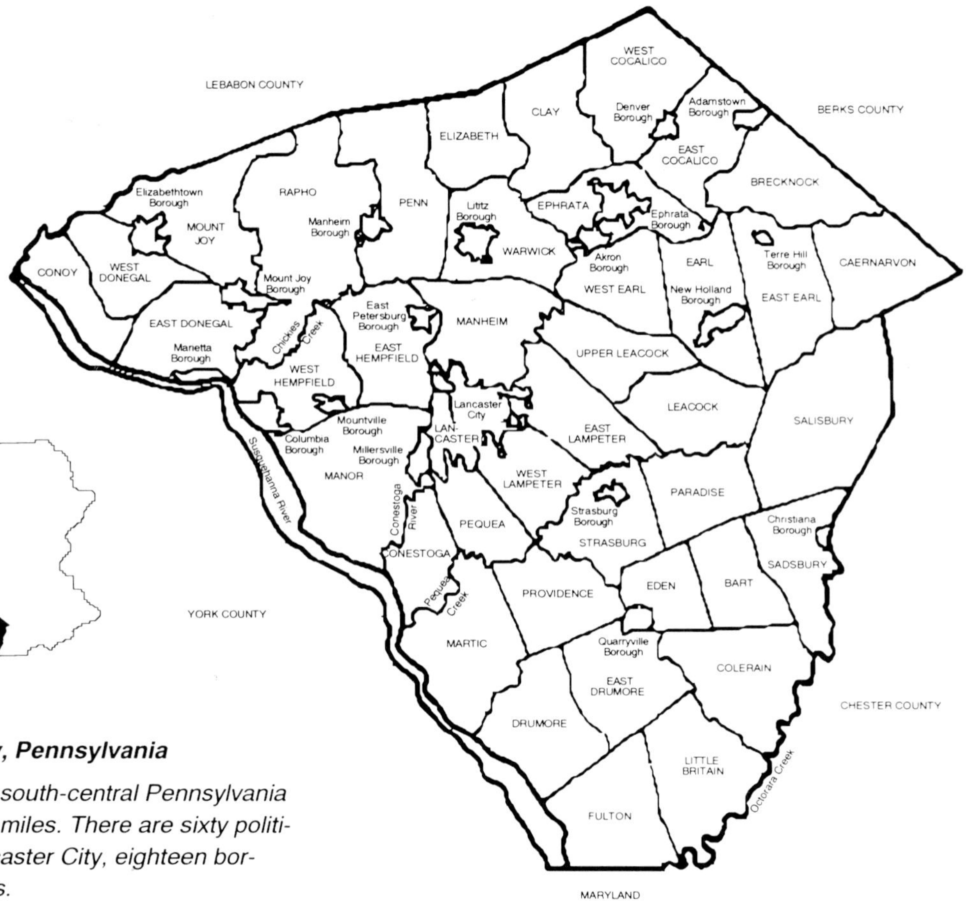

Lancaster County, Pennsylvania

Lancaster County is located in south-central Pennsylvania and encompasses 941 square miles. There are sixty political subdivisions including Lancaster City, eighteen boroughs, and forty-one townships.

Setting

Located in south-central Pennsylvania, Lancaster County is bordered to the south by Maryland, to the southeast by Chester County, to the northeast by Berks County, to the northwest by Lebanon and Dauphin Counties, and to the west by the Susquehanna River and York County. The county currently encompasses 941 square miles. Approximately seventy percent of its 602,240 acres remain in agricultural use.

Located to the southeast of Pennsylvania's Appalachian mountain ridges, within the Appalachian Piedmont, Lancaster County is relatively flat, largely consisting of an undulating plain averaging no higher than four hundred-fifty feet above sea level. The rock strata underlying the county generally divides it into three areas, each extending laterally across the county. The largest of these is a central limestone belt which spreads through the middle section of the county. While relatively flat, the limestone plain is broken by the Welsh mountains to the east and by Chickies Rock to the west, with a series of smaller ridges between. Three major waterways, the Conestoga River, Pequea Creek, and Chickies Creek, traverse the region, flowing southwesterly to the Susquehanna River. The northern tier is underlain primarily with sandstone and shale and includes a series of hills and ridges, with elevations reaching twelve hundred feet along the very northern tip of the county. Crystalline rock underlies the southern region of the county, which "has a wonderfully varied and picturesque maze of low hills and shallow winding valleys."[3] Waterways in the southern tier tend to lie within steep valleys and include the Octorara, Conowingo, Tucquan, and Muddy Creeks.

Weather in the county is generally mild. Soils are quite rich. Hagerstown loam covers most of the limestone valley; Penn and Berks soils, the northern tier; and Conestoga loam, most of the southern end. Rain is abundant and couples with the soil conditions to supply springs and wells. John Fraser Hart, in *The Land That Feeds Us*, states: "The only extensive areas of good farming land on the eastern seaboard of the United States are the limestone plain around Lancaster, Pennsylvania, and the Great Valley northeast and southwest of Harrisburg...."[4] This natural fertility has combined with Lancaster County's gentle topography and the skilled farming practices of local farmers to make it one of the most productive agricultural regions in the nation.

The pattern of human development on this landscape includes Lancaster City, a centralized urban area that has traditionally served as the county's hub for government, trade, commerce, and the arts. Most of the county's major transportation routes, including historic canal and rail routes, pass through or near the city. Along these routes, a series of smaller communities developed as regional market and trade centers. Until recent decades, virtually all of the county's remaining developed lands were in agricultural use. Today this historical pattern remains very much in evidence although suburban and industrial development have encroached on the farmland.

The county's population grew steadily from 3,485 in 1729 to 12,800 by 1738. Further increases averaged 10,000 per decade through 1920. During this time, the percentage of the rural to total population declined from one hundred percent to slightly more than fifty percent although the decline was minimal until about 1840. There are presently sixty political subdivisions, including the City of Lancaster, eighteen boroughs, and forty-one townships. The population in 1990 was 422,822.

• *Lori Stahl*

Settlement and Self Sufficient Farming

Circa 1710 - 1790

William Penn was granted title to Pennsylvania in 1681. It was his intention to create a colony of small landowners rather than a series of large estates held by a privileged class, as was common in other colonies. Together with Penn's belief in religious tolerance, Pennsylvania became a destination for Europeans seeking both economic opportunity and religious freedom.

When Penn acquired Pennsylvania, its Native American population stood at around fifteen thousand. By 1790, only thirteen hundred remained. In present-day Lancaster County, occupations by the Susquehannock, Conestoga, Pequea, and Shawnee have been identified. The century prior to 1700 saw prolonged periods of warfare between the Susquehannocks and the Five Nations of the Iroquois to control increasing fur trade. As a result of these conflicts, most of the county's Native American population abandoned the area by the end of the 1600s. The relative absence of Native Americans was an attraction for early European settlers wary of discord with existing populations.[5]

Yet the Native Americans left their mark on the landscape. Stevenson Fletcher, in *Pennsylvania Agriculture and Country Life, 1640-1840*, describes what the early European settlers often found: "Small cleared areas near streams were mostly the sites of abandoned Native American villages. Large cleared areas had resulted from periodic firing of the woods by Native Americans in pursuit of game; these were particularly useful to the first settlers of Lancaster and York counties." Native Americans also cleared out such areas for planting crops. Called *grubenland*, from the German word for small tree, the "Pioneer farmers of [Lancaster County] harvested their first crops of hay on these natural meadows."[6]

Prior to the creation of Lancaster County from Chester County in 1729, Pennsylvania was divided into three large counties: Bucks, Philadelphia, and Chester. The first Europeans settlers in present day Lancaster County were primarily traders who established posts along the Susquehanna River. The first grants of land occurred as early as 1691, although they were primarily speculative in nature, with owners rarely settling on the land. One thousand acres, called the "Servants' Tract," was granted to a number of Welsh servants who had reached the end of their obligation. They chose to return to England instead, and the land reverted to its Philadelphia owner in 1702.

Settlement

Substantial permanent settlement began to occur in what is now Lancaster County in the first decades of the 1700s. The first settlers were primarily Swiss and German Mennonites in search of a stable environment of religious freedom and economic opportunity. The sect was named for their leader, Menno Simons, who founded the Mennonite movement within the persecuted and fragmented Anabaptist sects of mid sixteenth-century Switzerland.[7] The Mennonites migrated into the Palatinate region along the Rhine River in present day Germany. There they experienced alternating periods of acceptance and discrimination as political and religious leadership changed.

A group of Mennonite settlers left the Palatinate and eventually settled in 1709 on a ten thousand acre tract along the north side of the Pequea Creek, south of the present day Lancaster City. Among these first settlers were Hans and Martin Meylin, Hans Herr, Martin Kendig, Jacob Miller, Martin Oberholtzer, and Wendell Bowman, progenitors of some of the county's most recognizable families. In subsequent years, other groups of Mennonite settlers arrived in Lancaster County, often assisted by the Committee on Foreign Needs, a society founded in Holland to aid Mennonite refugees. Another group of Mennonite settlers, numbering more than three hundred, arrived in 1717 from Manheim, where they had relocated after a particularly fierce round of persecution in their native Switzerland in 1710. Mennonite migration into Lancaster County continued throughout the eighteenth century.[8]

The Mennonites were followed in 1712 by a small group of French Huguenots who settled

• Ed Worteck

in the Pequea Valley. Like the Mennonites, the Huguenots came to this country after having suffered religious intolerance in Europe. With the revocation of the Edict of Nantes in 1685, King Louis XIV began to persecute Protestants living in France. Many migrated to present day Germany, settling predominantly in the Palatinate. Responding to the economic and religious opportunities of America, the Huguenots established footholds from South Carolina to New York by the turn of the eighteenth century. The Huguenot settlements in Lancaster began with the migration of Daniel Fierre, his mother Madam Fierre, and his brother-in-law Isaac LeFevre from the Palatinate to England in 1708. In 1709, they left England for America, settling first in New York and then arriving in Lancaster in 1712. They created a settlement on a 2,300 acre tract purchased by Madam Fierre from 3,000 acres that had been acquired by Martin Kendig in 1711, which he named Strasburg.[9]

By 1714, 58,937 acres of land in Lancaster County had been surveyed.[10] Scotch-Irish began to arrive in the area at roughly this time, and they located in the western portion of the county near the Chickies Creek, in what is now Donegal Township, and later along the Octorara Creek, in the southern portion of the county, from Sadsbury to the Susquehanna River. At about the same time, English Quakers began to settle in the far eastern portion of the county in what are now Salisbury and Sadsbury Townships. Present day Caernarvon Township was settled by Welsh Episcopalians in 1717.[11]

A group of Dunkers, a sect formed in Germany in 1708, arrived in Lancaster County in 1719 and established a church at Conestoga in

• Ed Worteck

1723. It was from this group that Conrad Beisel split in 1729 to form the Seventh Day Baptist Brethren at the Ephrata Cloister. A contingent of Amish settlers moved into the county by 1737, when a settlement was established northeast of present day Lancaster City.[12] Members of a 1732 Berks County Amish settlement relocated to eastern Lancaster County between 1754 and 1790. In addition, non-Mennonite German immigrants began to arrive in the 1720s, representing a variety of religious beliefs, including Lutherans, Calvinists, and Baptist Brethren.

Clearing the Land and Early Farming

Much of Lancaster County was covered by dense forest when its earliest settlers arrived. In their 1883 history of the county, Ellis and Evans, "according to the most reliable accounts," described the landscape this way: "heavy timber upon most of the limestone land, with occasional meadows and swamps; lighter woods covering the southern belt, shale and sandstone ridges, while some exposed areas may have been entirely barren of trees and verdure."[13] A variety of factors entered into the choice of land by early settlers. They may have preferred land resembling that from which they came, yet decisions were also influenced by existing settlements, a desire to settle in groups, etc. The largest portion of the limestone plain was settled by Pennsylvania Germans, with the Scotch-Irish settling in the Donegal area. The hillier southern portion of the county was largely settled by the English and Scotch-Irish.

Clearing the land and providing shelter for both families and livestock became the settlers' first tasks. During this early period, farming was practiced primarily to sustain the farm family. The earliest dwellings tended to be relatively temporary, providing shelter while land was cleared and crops were established. Often permanent barns were erected prior to the completion of a permanent dwelling. Houses constructed of logs were common in the first third of the 1700s and, as settlements became more firmly established, more substantial log or stone dwellings were erected. Among the earliest surviving dwellings in Lancaster County from this period are the log Mathias Slaymaker House (circa 1710) and the stone Hans (or Christian) Herr House (1719; National Register).

Mills for processing agricultural products were established along the county's creeks and rivers. Christian Schegel built a mill on the Conestoga about 1714 and others soon followed, including John Galbraith's in Donegal (1720), John Gardner's hemp mill on Chickies Creek (1721-1722), and Theodorus Eby's on Mill Creek by 1726. The mills were important centers of trade and meeting in this early period and were soon connected to surrounding farmland by paths and crude roadways. These trails would later serve as the basis for many of the county's permanent roads.

Farming Practices

With the increased availability of land, farming patterns developed in a markedly different way than had been common in Europe, where farmers often resided in villages surrounded by farmland on which "three field farming" or *Dreifeldwirtschaft* took place. In this system, crops were rotated on a three year cycle of "food, feed, and fallow," where, as Ivan Glick has described it, "one field would be idled in fallow, one field was planted in winter grain, and the third field was planted in spring grain or roots."[14] With the use of this system and because the generation of animal manure for fertilizer was limited by insufficient land on which to produce large quantities of hay, the soil was gradually depleted. When Mennonite and other refugees arrived in the Palatinate, they were forced to settle on depleted lands. Through experimentation, they soon discovered new feeds which could yield sufficient food to support their cattle on lesser acreages. Increased herds yielded greater quantities of manure. Manure nourished the soils and provided greater yields. To these techniques, they added the discovery of the use of gypsum and lime and the rotation of hay and other crops with legumes, both of which substantially replenished the soil. These practices made it possible to farm the land without depleting the soils while increasing its productivity.

About the same time, English farmers also discovered the positive impact of clover in replenishing the soil although the discovery was made for very different reasons. Sheep production became widespread as a result of increasing markets for cloth. Farmland was soon given over to pasture on which clover was grown in addition to grass. This rotation of grazing crops substantially improved the soil and, after a few seasons, made the land productive for other crops.

Both the Germans and the English settlers of North America brought knowledge of these renourishing techniques with them. Yet, for reasons that are not well documented, the use of soil conservation methods, including the spreading of lime/gypsum and the use of legumes within crop rotations, was not universal among settlers in this country. Several published historical accounts suggest that the vast quantities of land available in America made it easier for farmers to justify using crop rotations

• Lori Stahl

• Ed Worteck

that relied on a fallow period, or simply to move on to more productive lands, rather than practices involving gypsum and legumes. Fletcher suggests that the "introduction into grain farming of soil conserving crops, especially red clover and grass, was advocated and practiced by a few farmers for nearly a century before it was generally adopted."[15] There is evidence that the German farmers in Lancaster County were among the first in America to utilize these practices and to use lime.

Lime, a key ingredient for soil conservation and also an important building material, was extensively produced in Lancaster County by the mid 1700s. In 1754, following a visit to "the beautiful Valley of the Pequea," Governor Pownall found "on every farm a lime kiln and the land adapted for the best of wheat."[16] Limestone was burned in lime kilns to produce lime. Many of these structures remain scattered about the countryside.

In addition to lime, Lancaster County's German farmers were among the first in Pennsylvania to widely store and apply manure. Typically not practiced in the earliest settlements because of crude barn facilities, manure collection became common in local farms as soon as permanent barns were established.

Ellis and Evans described the early farming practices of the settlers:

The early farmers cultivated spelt, barley, oats, and buckwheat for summer crops, and rye for a winter crop. Wheat was not then extensively raised: it was considered too delicate a growth and uncertain yield...Flax and hemp were also cultivated, and manufactured into linen and wearing apparel....The fields were plowed in lands, a certain number of furrows being thrown together, leaving two uncovered furrows between the lands, as is still done in wet soil. In harvesting, two reapers would take a land. Young women who could be spared in the house helped, many of whom did a full day's work....The meadows were mowed earlier than the uplands and the hay dried by spreading and turning it in the field during fair weather, and by putting it on weathercocks over night or at the approach of rain....Corn was in that day, and even in later years, often topped and bladed, leaving the ears to ripen on the stalk....Threshing was done with a flail or by tramping it out by horses, - a labor that lasted through the greater part of the winter, and was cold and tedious work.[17]

Cows and sheep were added to the farms as early as it was practical to do so and provided dairy products and fleece. Large draft horses were also raised. Raising cattle for meat production was limited due to the abundance of wild game.

During the decades following 1710, particularly after 1725, the county developed rapidly. By 1729, Lancaster's population had risen to 3,485. In 1728, local residents petitioned to have a county created from Chester County. They were successful. On 10 May 1729, Lancaster County, named after Lancashire, England, was erected. At the time, the county included much of the western Pennsylvania frontier, only to be reduced to its present size through the creation of York, Cumberland, Berks, Northumberland, Dauphin, and Lebanon Counties between 1749 and 1813. Lancaster City was laid out by James Hamilton and became the county seat in 1730.

During the first third of the 1700s, farms began to produce greater yields and markets developed in Lancaster City and other population centers within the county. As quoted in *Fletcher's Pennsylvania Agriculture and Country Life 1640-1840*, Robert Proud stated in 1721: "The settlements about the Indian village of Conestoga [Lancaster] were considerably advanced in improvements, the land thereabout being exceedingly rich...They raise great quantities of wheat, barley, flax, and hemp...."[18] The marketing of crops beyond the county's borders was limited at the time by the difficulty in transporting goods. Completion of the King's Highway to Lancaster from Philadelphia in 1733 greatly increased such traffic. Trade was also established with Baltimore by the 1740s.

• *Ed Worteck*

Pennsylvania became the greatest producer of wheat in America by the 1730s, earning it the title "granary to the colonies." Wheat production was favored both because it was easily grown in the region and because it could be exported to the other colonies and abroad. By 1740, Pennsylvania flour and other grain products were being extensively exported to Europe. Lancaster soon dominated the colonies in wheat production. During the years between 1740 and 1790, wheat was responsible for much of the county's wealth. This was represented in local farms by the construction of larger and more substantial houses and barns. Seventy percent of the state's export trade was in wheat by the 1770s. By 1781, the assessed value of Lancaster County land reached $700,000.

Many chroniclers of Pennsylvania have given full credit to Lancaster's soil characteristics as the reason for its productivity. In reality, the practices of its farmers were a major contributing factor. During this period, Lancaster County's German population dominated the farming landscape. In terms of population, the percentage of the county's German speaking people fluctuated: in 1722 it stood at sixty-five percent; in 1758-1759, at fifty-eight percent; and in 1782, at sixty-eight percent.[19] These statistics and historical accounts reflect that some of the county's English settlers moved away, and often German farmers acquired their lands.

As Fletcher noted: "Former non-German settlements in Lancaster County are now evidenced by defunct or struggling Quaker meeting-houses and Episcopal or Presbyterian churches."[20] This shift has often been oversimplified as a testimony to the superiority of Pennsylvania German farming practices over those of the more transient English, Scotch-Irish, and Welsh. As early as 1789, Benjamin Rush wrote, "the German farm was easily distinguished from those of others, by good fences, the extent of orchard, the fertility of the soil, productiveness of the fields, the luxuriance of the meadow."[21] Recent historians, most notably James T. Lemon through his book *The Best Poor Man's Country*, have challenged this widely held notion, arguing that there were a variety of other factors contributing to the out-migration of non-German farmers from south-central Pennsylvania.

Lancaster County retained a substantial non-German population throughout its history, and the contributions of these people to its agriculture, economy, and cultural development are significant. Yet, if for no other reason than their greater population, the patterns of farming in Lancaster County were most influenced by the Pennsylvania Germans. The "Farms in Berks County, PA" multiple property nomination to the National Register of Historic Places described the typical Pennsylvania German farm this way: "The excellence of the Pennsylvania Dutch [German] farms was no accident; they were the result of good judgment, hard work, and superior methods of farming. Not only did the Pennsylvania Dutch [German] play a major role in the agricultural prosperity of Pennsylvania, but they established a tradition of family farming that has endured to this day."

The productivity of Lancaster County's farmers was not limited to wheat. Corn became an important crop in Lancaster County soon after settlement, primarily as feed for livestock. Rye was also raised by the early settlers for its straw, which could be used for thatch, making baskets, and also for its cereal. Grasses were not grown initially, as settlers relied on natural grasses for the feeding of their livestock. By the mid eighteenth century, "artificial" grasses became popular. Fiber crops, including flax, and to a much lesser extent hemp, were grown extensively. Pennsylvania also led the colonies in their production. Other popular crops included potatoes and tobacco.

Prior to 1790, livestock was raised principally for domestic use in Pennsylvania. Lancaster County's German farmers were noted for their livestock practices, whereby they treated and fed their cattle well and derived good yields of work and products from them. However, cattle of the settlement period typically produced little milk and were not extensively used to provide meat.

Chickens were raised primarily for domestic use in the years prior to 1790. While horses were not used extensively on farms in the colo-

• Ed Worteck

ny prior to 1790, they were used extensively to haul farm products to market. Lancaster farmers are credited with the development of a superior breed of draft horses by 1750. Fletcher states:

It was natural that the first great horse of American origin should be developed by the Pennsylvania Germans....Hitched to the great, canvas-topped Conestoga wagons - 'the ships of inland commerce,' as Dr. Benjamin Rush called them, they moved the heavy tonnage from farm produce of Lancaster, York, and adjacent counties to the Philadelphia market....It [was] a symbol of the transition period in Pennsylvania agriculture - between the pack-horse train and the railroad; between subsistence farming and commercial farming.[22]

Like livestock, the growing of fruit, vegetables, flowers, and ornamental trees was primarily a domestic activity in the colonial period. Apples were grown extensively and were the primary fruit used for domestic consumption on the farm, typically for the production of cider. Ciders and other fruit beverages, often fermented or distilled, were very popular among the county's German population.

Lancaster farmers experimented with the growing of silk in the 1700s. Susanna Wright, daughter of the founder of Wrightsville, is credited with having generated the county's best production with a mantua of more than sixty yards. Weaving was a major home industry in the county but gradually dwindled by 1790.

• *Ed Worteck*

Slavery existed in Pennsylvania, and Lancaster County's slave population in 1790, at 348, was third highest in the state. Slaves were mostly owned by the non-German farmers, as the Germans were generally opposed to the practice both for religious and practical reasons. Fletcher cites an exception in "James Kiemer, a German farmer in Lancaster County; he bequeathed eight slaves to his wife and children, with the stipulation that they be set free at specified times."[23] Pennsylvania passed America's first abolition law in 1780, which freed most slaves by the 1820s. Physical evidence of Lancaster's slave population is quite rare. A portion of a stone slave cabin remains within a barn at the Jasper Yeates farm in Salisbury Township.

Early Industrial Development

During this period, a number of industries developed in Lancaster County. Curtis Grubb, a Welsh settler, who established Cornwall Furnace in 1742, added extensive land in Lancaster County to his holdings, including Hopewell Forge which he acquired in 1765. Grubb's son Peter established Mount Hope Furnace in present day Rapho and Penn Townships in 1784. Windsor Forge, in present day Caernarvon Township, began producing iron in 1742. John Huber built Elizabeth Furnace about 1750, Thomas Smith began Martic Furnace in the southern end of the county in 1751-1752, and Old and Caldwell created Speedwell Forge on Hammer Creek in 1760.[24] "Baron" Henry William Stiegel established a glass factory in Manheim about 1765 and also acquired Elizabeth Furnace in 1757. Important eighteenth-century industrial developments attributed to Lancaster County are the Pennsylvania, or Lancaster, long rifle and the Conestoga wagon. As each of these industries developed, increasing numbers of Lancaster County's residents became involved in non-farming occupations, forming an expanding local market for local agricultural products.

The Farmstead, 1710-1790

During the county's earliest period of settlement, the farmstead generally consisted of a small farmhouse and barn. Shelter for the settler's family was always the first consideration although the first shelters were often quite temporary. Temporary shelter would be arranged for any livestock as soon as practical, and often the first permanent structure was the barn. Once a small portion of the total farm tract was cleared and crops were established, attention could be given to the erection of a more permanent dwelling. In Lancaster County, this typically meant the construction of a one story log cabin.[25] Outbuildings often were not constructed until the farm operation was more fully established.

By the mid 1700s, the full pattern of future Lancaster County farm complexes had been developed. The farmstead was dominated by the farmhouse and barn. A series of outbuildings associated with domestic functions were grouped about the farmhouse, and those associated with agricultural functions around the barn. Among the domestic outbuildings were kitchens (out-kitchens, summer kitchens), springhouses, bakeovens, washhouses, smokehouses, woodsheds, butcherhouses, and pumphouses. Agricultural outbuildings included pigpens, sheepfolds, chickenhouses, corncribs, hay barracks, wagon sheds, tool sheds, and often blacksmith forges, ice houses, and distilleries. Lime kilns and family burial grounds were other important built features of the farm. The building types common to the Pennsylvania German family farm dominated the county and were widely accepted among other cultural groups, although placement of buildings and styles and methods of construction varied.

By the later 1700s, new farmhouses constructed in Lancaster County tended to be two stories in height, generally three to five bays wide and two bays deep, and relatively conservative stylistically. During the mid to late 1700s, distinctions between Germanic and traditional English styles diminished as aspects of each were accepted by the other. The Georgian style began to have influence by the 1760s. In rural Lancaster County, examples of the style tended to be vernacular interpretations, with a few notable exceptions. It was also during the latter eighteenth century through the mid nineteenth century that the Pennsylvania barn reached its height of development. The size of barns increased. To meet changing needs for increased storage, additional threshing floors and upper level mows were accommodated.

• Ed Worteck

Commercial Farming
1790 - 1840

Commercial farming became an increasingly important aspect of Lancaster County agriculture throughout the mid to late eighteenth century. By 1800, Pennsylvania farmers turned the state into the leading producer of wheat in the country. Lancaster County farmers soon made it the leading wheat-growing county in the state. The county's wheat production led the nation on and off until 1850, when it produced 1,365,111 bushels and barely edged out Monroe County, New York. In addition, the acceptance of the land renourishing farming techniques of the "Agricultural Revolution" expanded rapidly in conjunction with the development of a major livestock industry. Improved transportation resulted in better access to markets and further stimulated Lancaster County's agricultural economy. Lancaster City was by this time the largest inland town in Pennsylvania with a population of 4,292. The assessed value of land in Lancaster County grew from $700,000 in 1781 to $6,700,000 in 1814. By 1830, it reached $28,700,000, four times the value of adjacent York County.[26]

The use of gypsum, or "land plaster," stimulated the growing of red clover which enriches the soil by adding nitrogen. The need to let fields lie fallow in order to control weeds was eliminated by the growing of corn, which is planted in rows, allowing farmers to hoe weeds from between the plants. Clover and corn became improved sources of feed for livestock. More livestock produced more manure, another vital source of nutrients for the soil. As John Fraser Hart described it in *The Land That Feeds Us*:

The new rotation greatly increased the amount of winter feed available for livestock....Farmers valued their animals for their manure almost as much as for their meat because they needed manure to fertilize their soil in the days before chemical fertilizer had been invented. You could tell how good a farmer was by the size of his manure pile. More manure meant richer soils, richer soils meant better crops, better crops meant larger animals, and larger animals meant still more manure. This interdependence of crops and livestock in a tightly integrated farming system is called mixed farming.[27]

• Ed Worteck

• Ed Worteck

The "Farms in Berks County, PA" multiple property nomination to the National Register of Historic Places described the importance of this period:

...the half-century from 1790 to 1840 has been called the golden age of Pennsylvania agriculture, a period when an agricultural economy reigned, before the age of industry brought its revolutionary changes to the patterns of work and life. This half-century was a time of agricultural awakening when worn-out fields were rejuvenated and farm mortgages paid. [The] new husbandry and new markets brought to Pennsylvania farmers a period of prosperity that has not been surpassed.

The development of Lancaster County as a major commercial producer of agricultural products was also aided by transportation and technological improvements, such as the cast iron plow, wheelbarrow sowers, and threshing machines. While Lancaster's farmers had been trading their goods in markets outside the county for quite some time, they had always been hampered by poor road conditions. Better roads and other forms of transportation allowed farmers to produce greater quantities of marketable products. As these markets opened, improved farm equipment greatly increased the farmer's capacity for production. The result was a sustained period of prosperity for local farmers that was soon reflected in their buildings and farms.

Ellis and Evans had this to say about the prosperity of the first half of the nineteenth century:

The orchards planted were beginning to yield fruit. There were barrels of cider on the shady side of the house, red and russet apples in the cellar, fragrant dried fruit, gathered seeds, and sweet herbs in the garret. The granary and the mows in the barn did not get altogether empty between harvests, The manure pile grew large and fat; and still there remained stacks of hay left over from last year. As the substantial comforts of the farmer thus multiplied other and new cravings and wants began to be felt, and he and his household needed more than ever the stern warnings from their ministers against the lusts of the eye and the pride of life....[28]

Transportation Improvements

The "Tobacco Buildings in Lancaster City" multiple property nomination to the National Register of Historic Places provides the following description of these transportation improvements:

Roads were laid out in what was to become Lancaster County beginning in 1718, connecting back country settlers along the Susquehanna and Conestoga Rivers to Philadelphia, New Castle on the Delaware, and ports in Maryland. Once the county was formed and the county seat located at Lancaster, a "King's Highway" was opened around 1730 which began at the old courthouse in Lancaster and ran almost to the division line of Lancaster and Chester Counties. For sixty years this "great road," or "Old Philadelphia Road," was the shortest way from Lancaster to Philadelphia. In 1792 the earliest turnpike in the United States was laid out between the (then) borough of Lancaster and Philadelphia. Known as the Lancaster Turnpike, it was opened in 1794; by the time Lancaster became the capital in 1799, the turnpike was heavily traveled by stagecoach. The construction of this advanced road was followed by others which linked the county to far-flung points around the state.

As part of Pennsylvania's internal improvements system, the Board of Canal Commissioners was appointed around 1826 to extend an existing canal program along the Susquehanna. In 1830, despite opposition, the canal along the river was extended southward to Columbia, a borough on the river about twelve miles west of Lancaster City. From Columbia, boats were either filled with produce and merchandise or the contents were transferred to carts and later railroad cars to be taken to Philadelphia. At Columbia large privately owned warehouses were erected to accommodate the canal boats. This commercial traffic soon was greatly expanded by rail and bridges linking Columbia with Pittsburgh: via this route goods and immigrants were shipped to the western states.

In 1820 the Conestoga Slack-Water Navigation Company was chartered with the objective of building several dams and locks between the mouth of [Conestoga Creek] and Lancaster for the navigation of boats; however, no progress was made on this until 1825. In 1829 large quantities of coal and boards began to arrive in Lancaster by this method. The navigation company alternated between prosperity and hard times for about thirty more years until it was finally displaced by the railroads.

The Pennsylvania Railroad Company was chartered in 1823; two of the incorporators were from Lancaster County. The purpose of the company was to build a railroad line between Columbia and Philadelphia, an ambitious project at this early date and one which predates the opening of the canals. This line was to divert the growing trade down the Susquehanna from continuing on to Baltimore. Although the plan at this time did not come to fruition, in 1826 another rail line was chartered, the Columbia, Lancaster, and Philadelphia. The major decision to building the expensive link between Columbia and Philadelphia, either by canal or rail, fell to the legislature which in 1828 passed an act authorizing the canal commissioners, then in charge of internal improvements, to extend what was known as the Eastern Division of the Pennsylvania Canal to the mouth of the Conestoga Creek and to determine the best means of establishing a route between Columbia and Philadelphia. Accordingly, the commissioners voted in favor of a rail line to be known as the Columbia and Philadelphia Railroad (later the Pennsylvania Railroad).

Engineering for the new railroad link between Columbia and Philadelphia began in May 1828 and was completed by December. Construction contracts were let in January 1829. The first passenger coaches were drawn by horses between Columbia and Lancaster on 31 March 1834 and the line from Columbia to Philadelphia was dedicated on 15 April 1834. Use of the railroad for transporting agricultural products took time to develop. The railroad did not replace the droving of cattle, for example, until after 1840.

• Ed Worteck

Crops

Wheat continued to be the county's major crop. Ellis and Evans described the crop rotation typical of the period: "Two years in grass, of mixed clover and timothy, one year in corn, one year in oats, then a heavy coat of barnyard manure plowed down, and two years in wheat, to be followed again by grass. Accordingly, many farms were divided into six fields, receiving one coat of manure every sixth year, and an application of lime once in the same year...."[29] Portable threshing machines were introduced into the county by William Kilpatrick around 1832. This was a substantial development as threshing, which previously was much of the farmer's winter work, could now be accomplished in a matter of days.

Corn became an increasing popular crop throughout this period. Native Americans introduced it to the European settlers, who raised it primarily for domestic consumption and as a feed for livestock. Because it was an important livestock feed, and because it grew well in Lancaster County's climate, the rise of cattle production resulted in an ever increasing demand for corn. This was furthered by improvements made in the way corn was grown after 1790, specifically the closer spacing of plants and the use of cultivators.

Rye also continued to be grown as a major crop and was used domestically for the baking of bread and the distilling of whiskey. Barley was produced primarily by German settlers for the production of beer. Production of hay rose due to increased demand brought on partly by transportation improvements that resulted in the greater use of horses. In 1828, the first leaf tobacco was grown in Lancaster County.

The use of gypsum, and its subsequent impact on the growing of red clover was a major development in the re-nourishing of the soil that became commonplace by the early 1800s. The increased production of grasses and red clover that resulted from the use of gypsum provided an improved source of feed for livestock. This resulted in larger herds. Manure from the herds was also used to renourish the soil. These developments, coupled with the expansion of livestock production and the better access to markets brought on by improved transportation, further stimulated Lancaster County's agricultural economy.

Historical accounts and agricultural statistics indicate that by the end of the eighteenth century, Lancaster County's farms were in a high state of cultivation. Duke De La Rochefoucault Liancourt, in his *Travels Though the United States of North America, The Country of the Iroquois, and Upper Canada*, published in London in 1799, described the farm of Colonel Matthais Slough, located just east of Lancaster, as he visited it in 1795:

This estate, which contains one hundred and ten acres, is now in a fine state of cultivation. About eighteen or twenty acres lie in grass, and form the most beautiful meadows; twenty-five are covered with wood, and the rest are under the plough. He lays from twelve to fourteen tons of dung on each acre: no land lies fallow, but he entertains the same prejudices as the rest of the farmers in favor of flat ridges, and against sheep. His son, in whose company I surveyed the estate, confessed, that the theory and practice which prevail in Europe do not agree with the husbandry of the Americans...." [30]

Cattle and Livestock

The period after 1790 also saw the rise of cattle and livestock industry in Lancaster County. With improvements in transportation, and the increased productivity of the land and its ability to provide surpluses beyond that necessary to sustain the farm family, the production and feeding of livestock became commonplace. Significant improvements in livestock husbandry resulted in greater productivity. Where the early settler's cow had often been a somewhat scrawny creature capable of producing only a limited quantity of milk, by the 1800s, cows were substantial animals of seven to eight hundred pounds. Locally grown beef increasingly became an important export commodity, and Lancaster County developed as a center for the fattening of cattle from the west prior to its shipment to markets in the east. According to John Fraser Hart:

Philadelphia and other seaboard cities were growing rapidly, and they needed ever larger quantities of meat. Their demands were satisfied, in part, by cattle from the newly settled areas of the west. Professional drovers assembled great herds of cattle in the back country and walked them eastward toward Philadelphia. The cattle lost weight en route, and they were lean, gaunt, and hungry by the time they reached Lancaster. The local farmers bought them cheaply, fattened them on corn, and shipped them off to the slaughterhouses of Philadelphia for a nice profit. The Lancaster plain became a major beef fattening area, and for many years Lancaster was one of the nation's leading cattle markets. [31]

Cattle were typically pastured in Lancaster County, with stall feeding practiced in the winter months. Cattle for sale in eastern markets were driven in herds in the late spring and summer. Early on, grass was the preferred feed. In later years grains, especially corn, were mixed in with the grasses. According to Fletcher, "Lancaster County, with its bountiful crops of grass and grain, became the center of the stall-fed beef industry of the state, a preeminence which it holds to this day [1950]."[32] Frank B. McClain, President of the Lancaster Livestock Exchange in 1924, reminisced about the early days of the cattle industry: "I remember well hearing my father tell the story of a drive of 300 head of steers he participated in, away back in 1843, from Champaign County, Illinois to Lancaster, and the progress made was about twelve miles per day on the average, and seventy days to the whole trip."[33]

Between 1790 and 1840, the commercial production of swine also developed as an aspect of Pennsylvania farms, including Lancaster County. Swine were relatively easy to raise and their meat could be cured for use throughout the year. Prior to 1790, swine that had been brought from Europe by early settlers existed in Lancaster. Rather than being held captive and fed, farmers routinely let them forage for food in their nearby woods. The swine of this early period were a sturdy breed of razorbacks. One early visitor was quoted as saying: "They have great stocks of Hogs kept in the Woods....I saw a Hogg Kill'd of about a Year old, which weighed Two Hundred weight; whose Flesh is much sweeter and even more luscious than that in England, because they feed and fatten on the rich (though wild) Fruits, besides those fattened at home by Peaches, Cherries, and Apples."[34] Between 1810 and the 1830s, competition from western swine producers, particularly in Ohio, increased significantly. Herds of hogs were driven overland across the Alleghenies to southeastern Pennsylvania. Although this affected commercial production in Lancaster County, local farmers continued to raise pork for their own use or to sell locally.

Sheep production, a major commercial enterprise in Pennsylvania, never developed significantly in Lancaster County. Again, competition from breeders in the west eventually led to the decline of the already minimal local production by about 1840. With the improvement of local transportation routes and the subsequent increase in the shipping of goods, the raising of horses became an important enterprise. The Conestoga draft horse continued to be raised until horsepower was largely replaced after World War I.

Dairy Production

Prior to 1830, the production of butter, cheese, and fluid milk was a secondary by-

• *Ed Worteck*

• Ed Worteck

product of the county's cattle industry. Most of the county's dairy production was consumed locally, either for on-farm use or as a market commodity in Lancaster City or the county's smaller boroughs. Butter was produced and stored in the cool of the springhouse. Commercial dairy production began to become an important industry with the increased demand for milk in the urbanized areas of the county, especially Lancaster City, brought about by their growing populations. Butter and cheese, however, were the most significant market dairy commodities prior to 1840.

Horticulture

The commercial production of fruits and vegetables increased significantly in the years following 1790, primarily as a result of increasing demand in urban areas and improved transportation. Apples, always an important product of Lancaster's orchards, were extensively grown and often converted and sold as cider. Another development on Lancaster's farms was the increasing occurrence of ornamental gardens. As farms developed and became more prosperous, the flower gardens that had always been a feature of the Pennsylvania German household gradually also became more elaborate.

The Farmstead

The major changes that occurred to the Lancaster County farmstead during this period were the increasing size and substantial character of the farmhouse and barn. H. M. J. Klein wrote: "...Lancaster county farms produce bigger crops per acre than the farms of any other county in Pennsylvania, and it is one of the reasons why the farmhouses and farm buildings of the 'man who follows the plow' in Lancaster fairly shout prosperity to the passerby."[35] Architecture saw a blending of cultural influences and an acceptance of elements of major national styles. By the 1790s, the Federal style of architecture became popular for local farmhouses, and the vernacular Pennsylvania style developed. Additional outbuildings began to appear as commercial farming developed and diversified. The regional "Pennsylvania" style evolved into a fully recognizable vernacular form of farmhouse and the Pennsylvania barn reached its full development.

Tobacco Production and Industrial Development

1840-1900

Lancaster County led the United States in total farm production in every U.S. Census from 1850 to 1900. In 1870, Lancaster County's total farm production reached $11,815,008, making it probably the first agricultural community in the United States to exceed ten million dollars. The county's 7,411 farms yielded 2,077,413 bushels of wheat, 2,820,843 bushels of corn, and 88,245 bushels of oats. The county led the state in the following categories: improved acres of farmland, cash value of farms, cash value of farm implements, total farm wages, total value of all farm production, value of animals slaughtered or sold for slaughter, value of livestock, number of horses, number of mules, number of swine, and in production of wheat, corn, oats, seed grass, and hemp. It was a close second to Chester County in orchard production.

The period from 1840 through 1900 saw many changes in the county's farming industry. Frederic Klein stated of the period from 1850 to 1860 alone: "In three distinct ways the Lancaster farmers of 1860 differed markedly from those of 1850. By the end of the decade they had become mechanized, organized, and specialized; mechanized through newly invented tools, organized through agricultural societies, and specialized by the rapid introduction of tobacco culture."[36]

In addition, the face of the county itself was changing as non-farm industries developed throughout the latter part of the century. Mostly centered in urban areas, they offered an alternative to farming employment for many Lancastrians. During this period, the City of Lancaster grew rapidly as a result of this industrialization and became an even larger market for local farmers. By 1870, the county's 1,616 manufacturing establishments ranked second in state and, by the 1890s, their economic impact began to rival that of its farms.

• Ed Worteck

From Wheat to Tobacco

In 1850, Lancaster County led the nation's counties in total wheat production by raising 1,365,111 bushels. Lancaster County continued to lead Pennsylvania with crops of 2,125,722 bushels in 1860, 2,077,413 bushels in 1870, and 2,232,590 bushels in 1890. York County, the state's next highest producer in 1870 raised 1,129,500 bushels, only slightly more than half Lancaster's total.

Yet by mid century, farming counties in the west were providing an ever increasing supply of the nation's wheat. The opening of the Erie Canal and the improvement of steamboat and early railroad service provided cheap transportation for western producers, and they soon overtook Pennsylvania in wheat production.

As western wheat production increased, many local farmers chose to work to improve the yield of their wheatfields in order to compete. A local strain of wheat, referred to as Lancaster red, became the regional wheat of choice for a time. For other farmers, simply increasing production was not enough, and they turned to another cash crop, tobacco. By the mid 1860s, Lancaster County was leading the state in the production of tobacco. In essence, it anticipated its declining supremacy in wheat and replaced it by a similar leadership in tobacco.

Tobacco

In 1990, the Historic Preservation Trust of Lancaster County completed a multiple property nomination to the National Register of Historic Places titled "Tobacco Buildings in Lancaster City." Gloria O. Becker, Ph.D., served as the principal researcher and author. The following excerpts from that nomination detail the history of tobacco production in the county.

The Rise and Dominance of Tobacco

Tobacco was considered the crop of choice because it fit into the Lancaster County farmers' crop rotation system and the weather was generally favorable for its growth. Many farmers, however, feared the plant would deplete the soil. It was later found that with planned rotation and manuring the plant could be cultivated without ill effects to the rich soil. Somewhere between 1840 and 1850, Lancaster farmers became the top tobacco producers in the country and remained atop the industry for more than seventy years.

Most early efforts at tobacco cultivation in Pennsylvania had taken place in the Philadelphia area and along the Susquehanna River, generally in Lancaster and York Counties. Farmers in Lancaster County had grown tobacco which had evolved from the species nicotina tobacum as early as 1828 near Ephrata, mainly for local consumption. These early attempts in Lancaster and York Counties produced a tobacco described as "shoestring," i.e., heavy, black, and gummy. Nevertheless, Germans of Lancaster County did grow their own, hang it in the barn rafters to cure, and used it for chewing or smoking. The unused leaf, although generally not properly cured, was rolled into cigars known as "stogies" (named for the nearby Conestoga River), and sold at local stores.[37]

The problem in shifting from tobacco culture as a local effort to one with commercial possibilities was directly linked to the type of leaf grown in Pennsylvania. Despite William Penn's hopes for tobacco cultivation and the colonists' early forays into its production--fourteen cargoes of tobacco were shipped from the Province of Pennsylvania in 1689--the quality continued to be inferior to that of Maryland and Virginia until about 1837.[38]

The change to a superior product came in the 1830s with the introduction of a Cuban tobacco seed called Havana, which produced a broader leaf and a better quality tobacco for use in the manufacture of cigars. Benjamin Thomas of York is credited with experimenting with this leaf in 1837 and distributing it to his friends the following year. It was then favorably received when shown in the Philadelphia market. Thus began the Pennsylvania cigar tobacco industry.

Several varieties of cigar leaf tobacco were grown in Pennsylvania and in Lancaster County, but the primary types were the aforementioned Pennsylvania Havana Seed, used for binders (inner wrappers), and Pennsylvania Broadleaf or Seedleaf, developed to have a broader leaf and used primarily for cigar filler (chopped inner leaves), and sometimes for binders and wrappers (wrappers were the outer leaves).

• Ed Worteck

The growth in the size of the cigar-leaf tobacco production in Lancaster and York Counties is impressive: in 1840 Pennsylvania produced 225,018 pounds of which York County produced 162,748 pounds and Lancaster County produced 48,860. By 1860 over 3,000,000 pounds were produced in the state with Lancaster County the leader in production. In that year, Lancaster County produced 63% of the total cigar leaf; by 1900 it would total over 90% of the state's production (49,335,407 pounds) valued at eight million dollars.[39]

By 1883 Lancaster City was second only to New York City as the largest seedleaf market in the country. The seedleaf product for cigars coming out of Lancaster City at that time, in addition to large quantities of leaf used for filler, now also included a fine wrapper leaf which was "soft, pliant, silky...not light nor flimsy, but thin and tough, with veins so small as not to show above the level of the leaf and only a moderate amount of nicotine." It was said to be "handsome in appearance and of pleasant flavor," with a rich dark-brown color, and it burned with a white ash.[40]

The success of tobacco cultivation in Lancaster County has been credited to several factors, especially the rich soil, favorable weather, and practice of its farmers to manure the soil to protect it from depletion. Another contributing factor was the presence in the area of many farm families, particularly among the Amish and Mennonites, who, with their strong work ethic, provided a ready supply of workers almost year round for this labor-intensive crop.[41]

In addition to soil conditions and tobacco-growing skills, the curing, packing, and marketing of the leaf at the farm level are also important. Tobacco grown in Pennsylvania is stalk harvested, i. e., cut in the field and allowed to wilt in the sun for about an hour to prevent breakage when handled. Next, the butt end is speared on a lathe, hung on temporary portable racks known as "scaffolds" or "tobacco ladders," and then hung in ventilated barns to cure for a minimum of eight weeks. In the ventilated barn or curing shed the green ripened leaf, which is harvested full of moisture and is brittle and difficult to burn, is converted to the brown, easy to burn, aromatic, finished cigar tobacco by a process known as air curing (in contrast to flue curing). (Additional drying of tobacco for storage and to further develop flavor and aroma is referred to as fermentation or "sweating.") Both dry days and humid days are needed to cure the tobacco in the barn. After drying, the tobacco wilts and is brittle, so farmers then waited for humid days or moist snowy days to allow the tobacco to draw in some moisture before making the final preparations prior to sale. This took place originally in the farmer's stripping cellar, or room with an earthen floor. The moisture softened the leaf and made it pliable so it could be stripped from the stalk leaf by leaf, sorted into good or trash grades, and baled.

When the tobacco was ready to be examined for purchase, buyers and representatives of tobacco dealers and manufacturers from all over the country came to examine the farmer's tobacco. If a sale was made, the bales were weighed, and a check was issued to the grower. After the grower delivered the tobacco to the warehouse, further processing, curing (sweating), and repacking of the leaf were the responsibilities of the packer-dealer at the warehouse. Proper management of the sweat was vital to the finish of the leaf; therefore, it was far better done in the controlled environment of the warehouse than the barn. Eventually specialized tobacco warehouses were built for sweating (or resweating, as it was called).[42] The development of this warehousing system to handle and store tobacco in the second half of the nineteenth century was directly linked to the growth of the tobacco products manufacturing industry, primarily cigar making, in Lancaster City.

The following comparison between Lancaster County tobacco production and that of the State of Pennsylvania is included in H. M. J. Klein's *Lancaster County Pennsylvania: A History*:[43]

Year.	State Pounds.	Lancaster Co. Pounds.	Lancaster Co. Per Cent.
1839	325,018	48,860	15.0
1849	912,651	378,050	41.4
1859	3,181,586	2,001,547	62.9
1869	3,467,539	2,692,584	77.7
1879	36,943,273	23,946,326	64.8
1889	28,956,247	19,217,800	66.4
1899	41,502,620	28,246,160	68.0
1909	46,164,800	36,892,868	79.9
1919	55,984,791	48,335,407	88.1

• *Ed Worteck*

• Ed Worteck

Cattle and Livestock

After 1840, Pennsylvania's dominance in cattle production declined as producers in Ohio and gradually further west assumed leadership. In Lancaster County, however, cattle and livestock continued to be a profitable aspect of total farm production. By 1870, the county had become the state's leading producer of "fat cattle." Fletcher quotes an 1879 description of the process by Frederick Watts of Cumberland County: "[Stock] is brought from the West, purchased in the fall, fed with hay, corn, and oats during the winter, and sold in the months of March, April, and May. The farmer finds that the increase of weight and the price of cattle fed is the most profitable mode of disposing of his corn and oats and increasing the quality and quantity of his manure."[44]

Prior to the third quarter of the nineteenth century, local cattle were driven on foot to locations as far west as Chicago and as far east as Philadelphia. As the freight capacity of the railroad increased, the county developed into a central processing point as stockyards developed along the railroads in and near Lancaster City. A small stockyard was established in Lancaster City in 1868 near the site of the present Armstrong World Industries plant. Two more opened near McGrann's Park within a few years. The Pennsylvania Railroad established the Union Stockyards in 1895 on a twenty acre site along the northern boundary of the city along Lititz Pike. At these stockyards, cattle and livestock would arrive by train from other locations across the United States and Canada, as well as from Lancaster County producers, and then be sold and shipped to others for processing.

In 1880, Pennsylvania was hit by an epidemic of bovine tuberculosis. The Pennsylvania barn was blamed for the higher incidence of the disease in Lancaster County because of the crowded conditions under which the cattle were kept. State agricultural experts predicted that all such barns would be demolished and replaced with new barns by 1900. Yet no evidence has been found to indicate whether or not any barns were actually demolished or modified as a result of these concerns.

During this period, the production of swine and horses continued to be significant aspects

• Ed Worteck

of Lancaster County's total agricultural production. Lancaster was among the leading producers of swine in Pennsylvania, topping all counties in 1860 with 54,826 and in 1870 with 50,070.[45] Sheep production continued to be relatively insignificant.

Dairy Production

According to Fletcher, "Between 1840 and 1900 it [the dairy industry] was transformed from a simple home enterprise, conducted mostly by farm women, into a highly organized commercial enterprise, conducted mostly by men."[46] The development of the railroad made possible the shipping of dairy products to markets outside Lancaster County after 1850. Prior to the development of electric refrigeration, such products could only be shipped to markets within a range of rapid delivery. Technological developments, including the invention of the vacuum condenser in 1856, the development of the silo about 1875, the invention of the continuous milk separator in 1879, and the development of a method of determining milk's fat content in 1892 contributed to the rise of the commercial dairy industry.

In addition, the breeding of dairy cattle underwent significant improvements during this period. Purebred lines were slow to gain acceptance by farmers, other than through experimentation by wealthy gentleman farmers, until the mid nineteenth century. Henry A. Carpenter was one of the earliest Lancaster County farmers to raise purebred cattle, acquiring a herd of one hundred such animals by 1830.[47]

Most milk was made into butter. Storage of the milk was accomplished in the springhouse,

• Ed Worteck

or in milk houses, where spring water would be used to provide a relatively constant refrigeration. Separation of the cream from the milk was accomplished using a system that included a milk pan (often made of earthenware or stoneware). Fletcher quotes an 1849 description of the process by a Lancaster County farmer:

> *...farmers here keep from 4 to 12 cows,...and keep milk in a cool spring-house in summer, and, during cold weather in a heated room; the cream is taken off as soon as it becomes sour, and kept in large crocks from 3 to 6 days; it is then churned, and the butter washed in cold water till the buttermilk is all out; it is then lightly salted...The farmers' wives usually take to market or exchange it for groceries as soon as made, and the country storekeepers resalt and pack it in small firkins, to be sent to other markets.*[48]

After 1883, centrifugal milk separators came into use.

Butter production gradually shifted away from the farm to the factory after 1880. Yet, in 1890, Lancaster County led Pennsylvania in the production of butter for the first time.

Market milk began to become a major dairy product for Pennsylvania farmers after 1870. Cities were growing rapidly at this time, and the market for processed milk was expanding. The rate of this growth can be seen in shipping from Lancaster County, which rose from 143,000 gallons in 1870 to 8,000,000 gallons in 1890.

The development of the silo was significant. Prior to this advance, farmers could only store dry feed for use during the winter months. Dairy cows fed dry feed produced significantly lower amounts of milk, and many farmers found it impractical to milk during the winter. Silos allowed moist feeds to be stored, thereby allowing for year-round dairying. Although invented in 1875, the first silo did not reach Pennsylvania until 1880. The round silo, so common in today's agricultural landscape, was not used extensively until 1890, and only became commonplace on Lancaster County farms about 1920.

Technological Advances

The period from 1840 through 1900 saw many significant technological advances and a rapid increase in farm mechanization. Between 1850 and 1860, the value of the county's farm machinery rose by more than fifty percent, to more than a million and a half dollars. Frederic Klein sates: "When we consider that the capital invested in all manufacturing enterprises of city and county in 1860 was only a little over five and a half million dollars, we can see the extremely significant part agricultural implements played in the economy of the region."[49]

The grain drill was first used in Lancaster County in 1842. The McCormick Reaper, which could harvest grains far more efficiently than by hand, was introduced into Lancaster County in 1851 by John B. Eby of Clay Township. The increased productivity of the reaper was augmented by the invention of a self-tying grain bundler.[50] Numerous other mechanical farm implements were developed during the period, and those in use were constantly improved. Jacob Mowrer of Lancaster County, for instance, patented a two row cultivator in 1869 that saw widespread use until the turn of the century.

In addition, the mid-nineteenth century saw significant advancement in the technology of plows. John Deere's invention of a plow with a circular metal saw blade in 1837 began what would become a series of major innovations in the design of metal plows. In 1859, Lancastrian Joseph F. Fawkes developed the steam plow, for which he won a gold medal from the United States Agricultural Society. Unfortunately, the invention was not well suited to the relatively hilly Lancaster County countryside and was primarily used on farms in the Midwest. By the 1860s, the use of metal plows was widespread in Lancaster County. Other developments, notably the riding plow, would follow by the turn of the century.

During this period, the use of horses to power farm machinery developed. "Horsepower rooms," in which horse labor was used to power belts and gears to power farm machinery, became additions to many of the county's barns. Samuel Pelton, Jr., a Lancaster County farmer, developed a horse powered threshing device in 1852. A remnant of a horsepower room remains at the mare barn of the Speedwell Stock in Elizabeth Township. The use of water and steam power also increased during the latter 1800s. The Rohrer's Farm and Mill in Paradise Township retains substantial vestiges of a water-powered system used to operate a sawmill, grist mill, and other farm machinery.

Yet in conservative Lancaster County, there was an appreciable level of concern over increasing mechanization. Day laborers especially were skeptical of new technological advances, fearing the loss of opportunities for work. There were also religious criticisms, with some literally interpreting the Bible passage "With the sweat of thy brow thou shalt eat thy bread."[51]

The diversification of Lancaster County agriculture continued throughout the latter nineteenth century. H. M. Engle of Marietta became a pioneer in the spraying of orchards in Lancaster County starting in 1889, when he sprayed his apple trees with London Purple for codling moth. He also became the first Pennsylvanian to experiment with the growing of chestnuts in the 1890s. Fruit production and truck farming continued to add to the county's total agricultural output.

Organization

The mid-1800s also saw the rise of farming organizations in Lancaster County. These groups served mostly as a venue for exchanging ideas and information about new methods and other practical topics. References to the Lancaster County Agricultural Society date to as early as 1852. In 1854, the Octorara Farm Club was established. It became one of the longest operating farm clubs in the nation.

During the 1880s, considerable tension developed between farmers and the railroads. Farmers generally resented the high rates the railroads charged them for moving their goods to market while charging the same rates to ship goods from other, more distant producers to the same markets. Pennsylvania farmers were also increasingly feeling the competition brought about by the settlement of the western United States, where between 1870 and 1880 more than 297,000 square miles of farmland were created.

The Farmstead, 1840 - 1900

Tobacco barns became commonplace during this period. New barns were also built to house livestock. Cribs, machine rooms, chicken sheds, and other outbuildings were added. Farmhouses were typically of the Pennsylvania style, often exhibiting simple Greek Revival detailing before about 1870. Buildings in a variety of Victorian styles were also constructed, as were Pennsylvania style farmhouses with Victorian details.

• Ed Worteck

• John P. Herr

Shifting Economies
1900-1945

Lancaster County did not lose its place as the nation's top agricultural producer until the early twentieth century, when heavily irrigated counties in the far west began to outproduce it. The Lancaster County Farmers Directory for 1909 listed the total agricultural wealth of Lancaster County as $20,767,146 and ranked first among the twelve most productive agricultural communities, ahead of second ranked McClellan County, Illinois, and third ranked Los Angeles County, California. The 1910 Census, however, ranked Lancaster second in total production behind Los Angeles, ending Lancaster County's seventy year agricultural supremacy.[52] By the 1930s, the county fell from the list of top ten producers, yet it continues to lead the nation's non-irrigated counties and more than four hundred thousand acres remain in cultivation.

Lancaster County continued to lead Pennsylvania in total agricultural production, notably in crop and animal production. In 1900 Lancaster produced $9,210,825 in revenues. Its 9,437 farms produced 28,246,160 pounds of tobacco; 1,906,440 bushels of wheat; 1,058,170 bushels of oats; and 4,523,550 bushels of corn.[53] 1906 saw the creation of the Lancaster County Farmers' Association, a cooperative established to help local farmers compete in the national market.

The years prior to World War I were generally prosperous for Lancaster's farmers. The county as a whole was experiencing tremen-

• *Ed Worteck*

dous industrial growth. Armstrong Cork moved to Lancaster from Pittsburgh and merged with several local cork manufacturers in 1908; the Stehli Silk Mill, one of the largest in the world, was in operation in Manheim Township; the Hamilton Watch Company grew steadily after its organization in 1892; the Follmer, Clogg, and Company Umbrella Works was the nation's largest umbrella manufacturing plant; and the Herr Manufacturing Company, makers of the Serta mattress, was founded in 1906. In addition, the county was a major candy-making center, produced shoes, had a wide variety of metalworking establishments, and everything from soap to bricks were made in its factories. Even with its increasing agricultural production, reaching $32,191,563 by 1919, the county's industrial production by this time had a far greater impact at $112,000,000. The number of farms reached 11,307 in that year.[54]

This increasing industrialization, in part spurred by the development of a trolley system throughout the county by 1910, would begin to impact the county's agriculture. One major trend that started during World War I and would become commonplace after the war was the decline in the availability and use of day labor on Lancaster County farms. Prior to the War, many farmers employed day laborers from the city or surrounding countryside. The war itself took many of these men away, and most who returned joined others who had left to take higher paying jobs in local factories. With the development of public transportation and the industrial growth the county experienced in the early twentieth century, industrial jobs gained increasing popularity. To some extent, the trend reversed itself after 1929, when the Depression put many of these same workers out of jobs and they returned to the farms. But the increasing mechanization of non-plain sect farms and the increasing population of plain sect farmers further reduced the demand for day laborers.

During the Great Depression, land values and market prices for agricultural goods dropped to the extent that some farms lost as much as half their value by the mid 1930s. Average farm income in Lancaster County dropped from three to four thousand dollars per year in 1927 to about eight hundred dollars in 1931.[55] Land values also declined, on average, from $9,052 per farm in 1930 to $6,782 by 1935.[56] Nature added its measure to these adverse conditions when the county suffered two severe droughts in the early 1930s. Sheriff's sales were a common occurrence.Later in the Depression, Lancaster's farms regained much of their vigor. 1937 saw one of Lancaster's most productive years in more than a decade.

Tobacco

Tobacco continued to be Lancaster County's most prominent crop between 1900 and 1945. According to "Tobacco Buildings in Lancaster City":

> *Between 1919 and 1940 the once-popular Pennsylvania Havana Seed, grown on the sand river bottoms and having a thin-bodied leaf when cured, declined substantially in importance. Pennsylvania Broadleaf or Seedleaf superseded it to become the primary type of tobacco grown in Lancaster County and the state. Lancaster County continued its dominance of Pennsylvania's tobacco production: in 1936 Pennsylvania produced 32,500,000 pounds of tobacco, of which Lancaster County produced 31,024,000 pounds or 96.4%, mainly the Pennsylvania Broadleaf or Seedleaf variety.*[57]

• *David B. Schneider*

Crops

The major development in crops in Lancaster County after 1900 was the growth in the production of corn as a feed crop for the county's increasing cattle and dairy production. Prior to 1900, corn was utilized primarily as a ripened grain or dry feed, and afterward, silage of corn steadily increased. Technological advances, among them the development of hybrid seed corn, resulted in substantially increased corn production by the 1920s: from 3,260,080 bushels in 1890 to 5,840,368 bushels in 1922. One of the county's major agricultural contributions occurred about 1911-1912 with the development of Lancaster Surecrop corn by Isaac Hershey, a Mennonite farmer. One of the three major lines of corn grown today is based on Lancaster Surecrop, which Hershey exported to farmers in other parts of the United States and North America. As a result of the increased production of corn, silos, which had been introduced in Lancaster County in the late 1800s, became commonplace after about 1920.[58]

Most Amish farmers in the early part of the twentieth century continued to practice a four crop rotation system, where hay was planted first, then corn, then tobacco in the following year, followed by wheat. Increasingly, cattle, dairying, and the production of chickens became more widespread.

Tobacco continued to be a leading cash crop throughout the twentieth century. Wheat production declined. Potatoes enjoyed a brief popularity among the county's farmers in the 1940s and 1950s.

Cattle

Lancaster County continued to be the largest cattle fattening center in the East, with cattle valued at more than $4,037,286 fed in 1910. Klein's 1924 history states: "In Lancaster County during the winter months there are made more pounds of beef from corn and other dried feeds, than in any other county in the United States." According to Klein, more than five thousand Lancaster County farmers "convert their corn, their hay and their fodder into beef from November 1 to June 1. Some of them feed thirty cattle, and a few as many as forty cattle, in one barn. But the greater number feed five or six head and the average, it is estimated, runs about nine head to each farmer." He further cited statistics from the 1922 Bureau of Statistics, Pennsylvania Department of Agriculture, showing 44,475 cattle with a value of $1,790,118.75. In 1922, 7,293 cars of cattle, 361 cars of hogs, twenty-two cars of sheep, and sixty-five cars of horses arrived and were sold at the Union Stockyards. Ten thousand head of Lancaster County cattle, worth $900,000, were shipped out through the stockyard that same year. In addition, Lancaster County farmers sold more than ten million pounds of hay and 6,500 bushels of corn to the stockyard to service the animals between the time of their arrival and departure. By the 1920s, more cattle were being winter-fed in Lancaster's stockyards than in any other location in America.[59]

Dairy Production

County-wide, production of dairy products increased substantially in the early twentieth century, even though local farmers experienced problems with bovine tuberculosis and its impact on milk markets in Lancaster and other cities. In 1893, efforts to control the disease in Pennsylvania began with the testing of herds. Lancaster farmers were generally opposed to mandatory testing from the outset. In 1926, the City of Lancaster passed an ordinance similar to ones passed by Philadelphia and New York that required milk sold locally come from animals that had been tested and found free of the disease. Lancaster County farmers reacted by forming the Farmers' Protective Association, the intent of which was to overturn Lancaster City's ordinance. A decision by the Pennsylvania Supreme Court in 1927 upheld Lancaster City's ordinance and the rights of the municipalities to regulate milk. Afterward, acceptance of the state's testing programs gradually became commonplace in the county. Notably, the last holdout from the testing program in Pennsylvania was a Lancaster County farmer, who complied in 1935.[60]

Lancaster milk producers found additional outlets for their product. The rise of the local

• Ed Worteck

candy industry provided an important one. In 1922, Lancaster produced 3,504,500 pounds of candy, consuming the equivalent production of milk from two thousand farms. By 1940, Lancaster County led the state in dairy production and ranked seventh nationally. In addition, Lancaster County led the state in butter production from 1900 through 1940.[61]

Poultry

Prior to 1900, poultry was typically a small industry for most Pennsylvania farms. Between 1900 and 1915, the industry boomed, due to sustained high prices and the availability of better breeds of chickens. Virtually all of Lancaster County's dairy farmers also began to produce chickens. Advances in poultry husbandry continued to result in higher and higher production. By 1940, the county led the nation in poultry production. Chicken hatching became a specialized industry after 1910 with the invention of the mammoth incubator.

With the vast increases in the number of chickens on each farm, and the increased incidence of disease, they were no longer allowed to roam about the farmsteads as earlier flocks had done when they were primarily kept for domestic use. By 1934, poultry houses, often reaching three or four stories in height, were common on farms throughout the county. Prior to 1920, chickens were primarily raised for egg production. Afterward, markets developed for poultry meat. Production of broilers became increasingly popular, especially after the development of the retail cut chicken trade after 1930. Accompanying the rise of chicken production was a corresponding increase in turkey production, which had all but ceased by 1910.[62] Both chickens and turkeys have remained important aspects of the county's agricultural production in the years since 1930.

The Stoltzfus farm in Leacock Township provides an example of the growing importance of poultry to the average local farmer. In the 1920s or 1930s, Jake Stoltzfus built a chicken house with a capacity of one hundred chickens. A subsequent owner of the farm, Christian Stoltzfus, added a second chicken house in the 1940s with a capacity of six hundred chickens. A twelve thousand cage chicken house was constructed in 1975.[63]

Lancaster County continued to be a leading producer of swine in Pennsylvania between 1900 and 1945, although its ranking had dropped to second behind York County by 1919.

Orchards and Truck Farming

Disease and insects began to become a significant problem in Lancaster County orchards after about 1920. By 1940, most small orchards were abandoned and replaced by a few commercial growers.

Increasing urbanization and industrialization resulted in truck farming, which became popular in the early part of the century, with small farms devoted to the raising of cash crops that could be sold from wagons, at market stands, or in the market houses of Lancaster City or the boroughs. Three major market houses, the Central, Southern, and Northern were popular in the city around the turn of the century in which producers could rent stands from which to sell their products.[64]

The Changing Farmstead

The changes in local agricultural production, notably the increases in cattle, dairy, and poultry, have resulted in both the adaptation of older structures and the construction of new ones. Cattle barns, additional dairy facilities, and chicken houses became commonplace. In addition, the increasing use of mechanized equipment, especially tractors and harvesting equipment since the 1930s, resulted in the need for adaptation of construction. Changes in regulations affecting the production of milk required farmers to construct separate milk houses for cooling and separating milk.

• *David B. Schneider*

Within the farmhouse, the introduction of electricity and modern appliances made summer kitchens and ground cellars largely obsolete by the mid century. Likewise, the introduction of indoor plumbing resulted in the virtual abandonment of the outhouse.

Suburbanization

Increasing population and improved public and private transportation began a trend toward the systematic conversion of farmland for residential development in the twentieth century. In 1890, the Lancaster Street Railway Company was formed to electrify existing horse drawn trolleys that served the City of Lancaster and its immediate surroundings. By 1891, an article in the *Lancaster New Era* appeared describing the advantages of suburban living: "Lots are cheaper in the outskirts than nearer the center, and larger grounds can be secured for the same money...for (most) some easy distance from the dust and the din of the busy city is more desirable as well as healthier than the turmoil and heat of the thickly settled portion." By 1900, thirty-five miles of track serviced the Lancaster City area with commuter service to Columbia and Lititz. Within ten years one hundred and fifty miles of suburban tracks connected the city with all of the county's population centers. 1900 also saw the introduction of the automobile into Lancaster County, when the Hershey Chocolate Company, which was founded in the county, purchased one.[65] With the subsequent popularization of the automobile in the early twentieth century, residential developments outside Lancaster City and the county's numerous boroughs and villages became increasingly popular. Early developments such as North Lancaster, West Lancaster, and Fairview, offered free trolley or automobile rides to prospective purchasers.

By 1930, non-farm property values began to rise relative to farm values. The average building value in Leacock Township, for instance, reached three-quarters of the average farm value. Farmland was becoming scarce and beginning to be overpriced for young farm families. The result was the beginning of the first appreciable out-migration of Lancaster County farmers, notably Amish and Mennonites, in search of land.

One non-farm development that did not materialize was the federal government's plan to acquire eleven thousand acres of land in the Donegal area on which to develop an explosives manufacturing facility. Frederic Klein provided this description of a meeting that occurred be-

• *Ed Worteck*

tween local farmers and the War Department in 1942: "the Lancaster County farmers, [spoke] slowly and softly about the generations of families who had made the soil of this area more fertile and productive than that of any county in the nation." The farmers were successful in convincing the War Department to locate the plant elsewhere.[66]

Private Estates and Suburban Landmarks

The suburbanization that occurred in the county in the first half of the twentieth century was accompanied by a trend toward the conversion of farms and farmsteads into country or suburban estates and suburban residences. The most locally recognized conversion occurred in 1927, when James Hale Steinman, a prominent local businessman and newspaper magnate, acquired the former Hershey Farm on Marietta Avenue from School Lane Hills Inc. and converted it into his mansion complex, called Conestoga House. The Steinmans utilized the services of noted local architect Frank J. Everts, assisted by nationally acclaimed artist Charles Demuth, to transform the traditional Lancaster County farmhouse into a Colonial Revival style mansion retaining much of the character of its nineteenth-century design. The complex was further refined between 1937 and 1940 when the Philadelphia firm of Mellor and Meigs designed the present formal gardens, outbuildings, stables, garages, and other exterior features.[67] Other farms have been converted for use as "gentleman" farms or country estates, where commercial agriculture is either no longer practiced or is secondary to the primarily residential or resort activities of the property. Within suburban developments, especially those created in the early twentieth century, farmhouses and occasionally their associated outbuildings, have often been retained as single family residences or as centers for community activities.

• David B. Schneider

A Changing Landscape
1945 - Present

The end of World War II marked a turning point in Lancaster County's history. The Post-War Planning Council, created in 1943, marked the beginning of planning for the expansion of urban facilities, principally water and sewer lines and roads, into the rural areas of the county. Increased demand for housing, fueled by federally-assisted mortgages from the Federal Housing Administration and the Veteran's Administration, resulted in rapid suburban development in the late 1940s and 1950s. Retail, banking, and transportation facilities in the center city were gradually disbursed throughout the county. According to the Lancaster County Planning Commission, "by 1954, there were more than 100 homes being built in Lancaster County each month."[68]

In 1940, Lancaster recorded 8,823 farms, the first time since the eighteenth century the number of farms in the county actually decreased. Since 1940, the downward trend continued steadily:

Number of Lancaster Farms 1945 - 1987

Year	# of Farms	Acres in Farms
1945	8,823	509,424
1950	7,952	495,500
1954	7,951	498,206
1959	7,043	482.579
1964	6,247	467,325
1969	5,323	426,100
1978	4,915	425,561
1987	4,775	403,964

Even with this decline, the county continued to lead the state in value of production: $335,180,200 in 1975 and ranked twelfth in the nation among all counties and first among non-irrigated counties.[69]

By 1975, livestock and dairy and poultry products accounted for almost three-quarters of the county's total agricultural output, with

• *David B. Schneider*

dairy products alone accounting for twenty-five percent of the total.[70]

In recent decades, growing suburbanization has had a dramatic impact on Lancaster County. More than ninety-two thousand acres of farmland have yielded to development since 1959. Since the late 1970s, the trend has accelerated. With one of Pennsylvania's fastest growing populations, the county lost fifteen thousand acres to development in 1987 and 1988 alone.[71] Since 1980, according to the Lancaster County Planning Commission: "approximately 4,800 acres of land have been approved for development every year. This translates into 68 square miles of land over a ten year period to accommodate approximately 60,000 people. In contrast, Lancaster City accommodates approximately 60,000 people on only seven square miles of land.[72] Since 1959, the county has lost an average of eight acres of farmland per day.[73]

Preservation and Community Planning

In his *Lancaster County 1841-1941*, Frederic S. Klein described remarks made by William Uhler Hensel before the Lancaster Board of Trade in 1900: "[If] Lancaster County made the same strides in the coming century that it had made in the past one, a hundred years hence we would have a population of 1,100,000, Columbia would be an annex and Philadelphia a suburb."[74] While the realization of Hensel's vision appears unlikely, the changes it envisioned for the county have become partially true. Population grew from about 320,000 in 1970 to 422,822 in 1990 and is predicted to exceed 540,000 by 2010. Lancaster and Columbia, while not annexed to one another, are rapidly being joined by suburban development. And rather than Philadelphia being Lancaster County's suburb, Lancaster County has become a home for commuter residents to and from metropolitan Philadelphia, Chester County, and Harrisburg.

In recent years, the Lancaster community has focused a great deal of attention on reversing the trend toward scattered, unplanned suburban sprawl. In 1975, the Lancaster County Planning Commission issued a comprehensive plan for the county that called for improved growth management to prevent the irretrievable loss of prime farmland and historic character.[75] The commission is currently developing a sweeping and innovative growth management strategy.

Farmland preservation has recently become both very successful and very popular and has included noteworthy private, nonprofit, and governmental activity. Lancaster County established the Lancaster County Agricultural Preserve Board in 1980. This governmental effort was supplemented in 1985 by the formation of a private, nonprofit organization to promote farmland preservation. In addition to its advocacy role, the organization, now known as Lancaster Farmland Trust, has also been actively pursuing permanent agricultural easements. Between the two entities, more than sixteen thousand acres were preserved through easements by 1993. In addition, most local municipalities now have some form of effective agricultural zoning to promote the continuance of farming. As a result of these efforts, there is an excellent chance that at least a sense of Lancaster County's traditional agricultural character will survive.

The Historic Preservation Trust of Lancaster County, founded in 1966 as a county-wide advocate for historic preservation, has in recent years focused greater attention on the rural cultural resources of the county. Through its Rural Preservation Project, under which this study has been completed, and by cooperating with elected officials, with the Lancaster County Planning Commission, with the Lancaster Farmland Trust, and with other interests, attention is being focused on the national significance of the county's agricultural heritage as it is expressed in the built environment.

2: The Agricultural Landscape

The landscape of Lancaster County is still dominated by farming. The farm is the basic unit from which the broad fabric of the county's agricultural landscape is made. In Lancaster County, farms generally exist within the context of active agricultural districts where farming activity is broken only by occasional alternative land uses along transportation routes or modern suburban residential development.

The pattern of large expanses of open cultivated fields interspersed with complexes that include substantial farmhouses, large barns, and numerous outbuildings has remained relatively unchanged since the 1700s. A number of written descriptions of Lancaster County survive from the eighteenth and nineteenth centuries. Several of them were edited by William Frederic Worner, Librarian of the Lancaster County Historical Society, in his 1927 book *Old Lancaster: Tales and Traditions*. While most of the descriptive accounts Worner uses in the book relate primarily to Lancaster City, the following give indication that travellers to the area have long recognized the richness of the county's agricultural landscape.

One of the earliest descriptions of Lancaster County in *Old Lancaster* is found in a series of excerpts from *Travels Through the Interior Parts of America in a Series of Letters By an Officer*, written by Thomas Anburey in 1789. Anburey

Jacob Miller Farm

A circa 1848 painting by Arthur Armstrong captures the essence of Lancaster County's agricultural landscape. Scenes and vistas such as this have been recorded since the 1700s. *• Courtesy Gerald Lestz*

was a British lieutenant who had served in America during the Revolution and who later wrote a two volume description of the country as he experienced it. Anburey's comments on Pennsylvania farming are not limited to Lancaster County, but undoubtedly reflect what he saw during his visit here in 1778:

After you get over the Delaware, a new country presents itself, extremely well cultivated and inhabited; the roads are lined with farm houses, some of which are near the road, and some at a little distance, and the space between the road and houses is taken up with fields and meadows; some of them are built of stone, two stories high, and covered with cedar shingles, but most of them are wooden, with crevices stopped with clay; the ovens are commonly built a little distance from the house, under a roof to secure them from the weather.[1]

Anburey continues: "The farmers of Pennsylvania, and in the Jerseys, pay more attention to the construction of their barns than their dwelling houses."[2]

In 1783, Johann David Schoepf, a German physician, visited Lancaster. He published a book on his travels in America in which he recorded his observations on Lancaster City and County. *Travels Through the Confederation* was published in Germany in 1788 and its 1911 English translation is quoted in *Old Lancaster*:

The road to Lancaster lies through a limestone valley, a fertile, varied, and well-farmed region. Along the road, indeed, ones sees, for the most part, sorry cabins, for the better houses of the well-to-do landowners are set a little off the road. This and the custom of always leaving some timber next to the road, brings it about that travellers think they are going through nothing but wilderness when all around them are plantations and dwellings stuck away in the bush. On this road everybody I met I addressed in German, and they all answered me in the same language. Very many Anabaptists live in these parts; good, kind people, and sturdy subjects who here, as well as in Germany, win the love of their neighbors and the regard of the magistracy.[3]

Unlike most of the verbal sketches of Lancaster surviving from the eighteenth century, which typically focus on the eastern portion of the county, Duke De La Rochefoucault Liancourt provides a description of the area northwest of the then Borough of Lancaster. In *Travels Through the United States of North America, The Country of the Iroquois, and Upper Canada*, published in 1799 and based on his visit here in 1795, is contained the following:

The road from Lancaster to this place [Maytown] lies chiefly through a woody tract of country, which assumes a wilder appearance than we have hitherto seen. Cultivated land appears more rarely as we proceed, except a few valleys, which still lie in grass, or are sown with Indian corn. In proportion as the distance from Lancaster increases, houses of brick or stone are less frequently seen. We met with scarcely anything but log houses; every where we observe German farms, small houses, and large barns. Cows and oxen, which seemed tolerably good, we found grazing in the woods and near the road; and also saw, at times, sheep, but never more than eight or ten of them together.[4]

Jedidiah Morse, father of Samuel F. B. Morse who invented the electro-magnetic telegraph, was himself a noted geographer who was often referred to as the "Father of American Geography." Morse's 1797 book, *The American Gazetteer*, includes a brief passage about Lancaster County: "Lancaster is a populous and wealthy county in the interior part of Pennsylvania....The lands in this county are rich and well cultivated. The hills in the northern part abound with iron ore..."[5]

Englishman Fortescue Cuming visited Lancaster in 1807 and his description of it was published in *Sketches of a Tour to the Western Country*...," published in Pittsburgh in 1810: "The road continued fine, and the country rich, laid out in large farms, with good dwelling houses of brick and stone, and immense barns."[6]

Worner includes a description of Lancaster County by Mrs. Anne Royall, whom he terms "America's pioneer woman journalist." Mrs. Royall produced a ten volume work entitled *Travel in the United States* in 1829 in which she related her impressions of the county from a visit here in 1828:

The beauty of Lancaster appears to just burst upon the traveler as he proceeds to Lancaster from Reading, and becomes more beautiful every mile of the distance. Two things suggested themselves to me as I drove this fertile state, viz. that neither the inhabitants themselves, nor the inhabitants of other states, have any idea of the superiority of Pennsylvania over other states, it being almost out of the common routine of fashionable tours; and no writer has, or perhaps can, do justice to Pennsylvania.

The country is visible to a great extent, the road sometimes rising upon gentle swells; the land then suddenly changes to a rich soil, the whole chequered with fields, meadows, orchards, and adorned with alternate patches of lofty timber...."

As we draw near Lancaster, it still grows more beautiful. Level as far as the eye can see, and as rich as any bottom land on the western waters, and in such a beautiful order, – the ploughing, the fences (post and railing), exact and regular to a mathematical nicety, – not a particle to mar the beauty of the prospect.[7]

Worner's selection of descriptions of Lancaster County ends with the visit of Mrs. Royall. In 1848, noted artist Arthur Armstrong painted a view of the 1805 Jacob Miller House and Farm in Lancaster Township. The painting shows a farm scene, complete with the farmhouse, bank barn, mill, and other outbuildings. Similar scenes are still prevalent throughout the county today. Another written account, dating from 1853, can be found in Eli Bowen's *The Pictorial Sketch-Book of Pennsylvania*:

As we leave Chester County, and pass through the range of hills called Mine Ridge, the great county of Lancaster, in all its glory, expands before the eye. An intelligent Englishman called this county the 'garden of America,' and a view of it from this position will fully justify the propriety of this compliment. It is, without a doubt, the garden of this glorious Union, and there are few spots in this wide, wide world, which could present a nobler scene to the eye than is here afforded....

The whole of the country is, therefore, in the highest state of cultivation; and in the economy which characterizes the general agricultural system, there is probably not a more prolific region in the United States.

Bowen continues his description of the farms themselves.

In passing through this splendid agricultural region, the stranger will particularly observe the neatness and order which characterize the general aspect of the scene of farming operations...the substantial and comfortable buildings, especially the imposing appearance of the barn. Nearly every large farm has a cluster of buildings, the most prominent of which is the barn, situated next to the mansion house, around which are scattered wagon and carriage-sheds, corn-cribs, wash-house, summer dining-house, etc., with adjacent tenant house.[8]

An engraving of the "General Appearance of a Lancaster County Farm" accompanies the above description and depicts a pastoral farm scene still common in the county today.

"General Appearance of a Lancaster County Farm"

Engraving from Eli Bowen's 1853 book The Pictorial Sketchbook of Pennsylvania.

Present in the drawing are a substantial farmhouse and barn, an outbuilding, and a mill. A series of similar, yet more detailed, engravings are included in Everts and Stewart's 1875 *Atlas of Lancaster County*. Again, the continuity of the pattern of farming activities on the landscape shown in these sketches remains evident in much of the present day county.

It is interesting to note that early writers acknowledged the prominence of Lancaster City as the centralized hub of the agricultural county. Dr. Schoepf described the city as he saw it in 1783: "Lancaster, of all inland towns of America, is the most considerable, numbering already 900 houses, and it is hardly 80 years since the place was first established." The Reverend Manesseh Culter said of the city after a visit in 1788: "Lancaster is a large and ancient town; the best built inland town in America." Jedidiah Morse, writing in 1795, called the city "a handsome and flourishing post-town, the capital of Lancaster county, Pennsylvania, and the largest inland town of the United States...." Anne Royall captured the essence of the town's relationship to the countryside: "Lancaster city is surrounded by a large unbroken body of the richest land in Pennsylvania, or in any other of the Atlantic states."[9]

The Patchwork Quilt

This aerial photograph of Lancaster County illustrates the expansive nature of Lancaster County's agricultural landscape and shows many of the elements that define its character. *• Lori Stahl*

3. Farm Buildings in Lancaster County

The entire history of farming in Lancaster County is reflected in the historic farm buildings and structures that remain on the landscape. The areas settled by the county's Germans, Welsh, and other groups can still be identified by studying the stylistic differences between their buildings. Likewise, evidence of changing cultural patterns can be seen. Abandoned Quaker meeting houses may reflect the moving on of English settlements, and the alterations made to existing farm buildings by Amish families exemplify this. In addition, natural features of the land influenced building construction. Log houses are still common, although rarely visible beneath layers of more recent construction, and reflect the abundance of timber found by early settlers. Limestone farmhouses, barns, and outbuildings are found within the limestone plain. Sandstone and brownstone construction are common along the northern tier of the county. Bank houses are common in many areas of the county and respond both to site conditions and functional requirements.

Stylistically, farm buildings in Lancaster County tend to be relatively conservative, although high style examples remain of all major architectural styles except Greek Revival. All farm buildings, both high style and vernacular, reflect high quality craftsmanship and construction detail. Many buildings show evidence of one or more stylistic or cultural influences.

Windom Mill Farm

The Windom Mill Farm is located in Manor Township. *• David B. Schneider*

For example, many of the county's Georgian and Federal style farmhouses retain Germanic features. This cross-acceptance of influences was in part responsible for the development of the vernacular Pennsylvania style farmhouse that began to appear in the early 1800s and to which details of later stylistic periods were applied. Likewise, the Pennsylvania barn, so much a symbol of the Pennsylvania Dutch culture region, represents an adaption of earlier forms and the acceptance of new technologies.

The cultural traditions of the county's plain sect farmers are also evident in the built environment. The importance of family is indicated by the construction of additions, including the "Dawdy Haus." The practicality of low maintenance and the restraint of ornament gives rise to the use of synthetic siding materials and replacement windows. Other structures are often added to farmsteads to accommodate differing technologies and operations from non-plain sect farms. For example, windmills remain common since connection to outside sources of power are not accepted. In Amish houses, where religious meetings are held, interior floorplans are open, and front porches are often enclosed to provide additional gathering space.

It is precisely this melding of cultural, stylistic, and technological influences and their adaptation to change that defines the distinctive character of Lancaster County. In many respects the county's built environment reflects the embodiment of the American ideal: a place where different cultures met, adjusted to one another, and forged a community, based on the toil of the common man rather than a reflection of its social elite. For these reasons, the commonplace, the vernacular aspects of Lancaster County's architecture, are particularly significant.

This continued adaptation to changing technologies and market conditions is also a major character-defining aspect of Lancaster County farms and farm buildings. In assessing integrity, or the degree to which something conveys a sense of time and place, it is essential that consideration be given to the significance of the continuing viability of the farm and its ability to adapt to future changes. The farmstead must be evaluated as a vital and changing complex, and the significance of individual components must be assessed for their contribution to the whole.

Martin Barr Farm

The Martin Barr Farm is located in Strasburg Township. *• David B. Schneider*

Adaptation to Change

Buildings are altered to meet changing family needs. • *Ed Worteck*

The Farm

For the purposes of this study, the definition of a farm contained in the "Farms in Berks County, PA" multiple property nomination to the National Register of Historic Places provides a useful description reflecting the nature of farms in Lancaster County:

The farm consists of land and land forms, buildings and structures utilized by man for the production of agricultural goods,....the farm is commonly a "family farm", operated by the members of a family working as a cooperative unit. The farm includes a combination of natural and cultural or man-made features, such as cropland, woodland, wetland and waterways with different topographical and soil characteristics, as well as fences, roads, lanes, bridges, lime kilns, walls, springs, ponds, contour strips, ditches, terraces and groups of buildings for domestic and agricultural use. Each farm is unique in its physical and cultural features, its history and its current use. Although some farms are representative of a period of architecture or a type of use, most farms exhibit a combination of periods and types, and an evolution of farming customs and practices.

Even though early settlement in Lancaster County was multi-cultural, the Pennsylvania German farm form serves as the primary model for the description of the farm in Lancaster County. Several important books have been written about farm buildings in this region. Among them are *The Pennsylvania German Family Farm* (1971) by Amos Long, *The Pennsylvania Culture Region: A View from the Barn* (1986) by Joseph Glass, and *The Pennsylvania Barn* (1991) by Robert F. Ensminger. The following descriptions of the various components of farms in Lancaster County have been developed from these studies, coupled with analysis of the Lancaster County Historic Sites Survey.

The Farmstead

On most Lancaster County farms, both domestic and agricultural buildings and structures are centralized in a relatively close-knit complex, here referred to as a farmstead. The location of the farmstead on the farm unit varies by farm and is based on topography, road access, and other considerations. A typical Lancaster County farmstead consists of a farmhouse, barn, other outbuildings serving both domestic and agricultural functions, yards, gardens, and roadways, paths and other circulation patterns. In Lancaster County, farmsteads are often not delineated from adjoining croplands, orchards, and occasional woodlots except by minimal fencing, ditches, or landscape features.

The principal buildings within a farmstead are the farmhouse and the barn. Outbuildings serving domestic functions, such as tenant houses, springhouses, kitchens, bakeovens, root cellars, distilleries, smokehouses, woodsheds, privies, washhouses, butcherhouses, and pumphouses are generally clustered around the farmhouse and its immediate yard. Farming-related outbuildings, including tobacco barns, hay barns, wagonsheds, carriage houses, toolsheds, pigpens, chicken houses, sheepfolds, corncribs, and milkhouses tend to be sited around the barn. Additional features, such as lime kilns and family burial grounds, are often located on the farmland, most often some distance from the farmstead.

Among the site characteristics that influenced the location of farms and the placement

A Farmstead

• Lori Stahl

of buildings were topography, orientation to transportation routes, access to roads and fields, and the availability of fresh water supplies. Sloping sites provided drainage of water away from buildings. Springs and streams provided an elementary form of refrigeration.

The number, size, and function of buildings varies with the size and type of farm and its evolution over time. Arrangement of buildings within the farmsteads is a product of the original owner's concepts of what would prove most productive, along with alterations to that layout, made by subsequent owners, and vary based upon site conditions. Symmetrical layouts, irregular ones as well as linear placements, are common. While barns are often the most substantial buildings within the farmstead, farmhouses are typically the visual focus. This is often accomplished by the placement of the house closer to the road, in a central location, on a higher elevation, etc. Barns and farmhouses are often sited with their roof ridges parallel to one another, although perpendicular, and angled placements are common.

In more recent times, a variety of additional farm structures, such as silos, manure handling and storage facilities, and equipment sheds have been added to meet changing conditions. The adaptation of Lancaster County farms to technological and market changes over the past two centuries has resulted in its continued present day economic viability. This is an integral part of their significance that is represented in the evolutionary character of most farms and their buildings.

Farm Scene *• Ed Worteck*

The Farmhouse

The farmhouse serves as the principal dwelling unit on the farm. Unlike the evolving function of other farm structures, the principal function of the farmhouse as a residence for the farm family has generally remained constant in Lancaster County, where most farms continue to be family farms.

In Lancaster County, farmhouses remain from all periods of the county's history. The earliest farmhouses tend to reflect relatively undiluted architectural influences of the ethnic groups that created them. However, many of the surviving examples have also been extensively altered. Among the earliest farmhouses recorded in the Lancaster County Historic Sites Survey are a number of log dwellings, many of which have been incorporated into larger evolutionary houses or concealed behind later finishes and architectural embellishments. Among the county's early log buildings retaining their historic exterior appearance is the Mathias Slaymaker House (circa 1710) in Paradise Township.

In 1981, the *Journal of the Lancaster County Historical Society* published Arthur C. Lord's analysis of the Direct Tax of 1798 as it applied to housing. The tax lists studied covered most of the western half of the county. According to Lord's findings, more than forty percent of the houses outside Lancaster City were constructed of log, approximately forty to forty-five percent were stone, ten to fifteen percent were brick, and five to ten percent were frame. More than seventy percent were one story in height, and their average size was less than seven hundred square feet. While these statistics seem to indicate the relative rarity of substantial farmhouses, the numbers of larger stone and brick houses in the nine present day townships covered by the study are impressive: 488 stone houses, 78 brick houses, and 536 multistory houses.[1]

Germanic

The Hans (or Christian) Herr House (1719; National Register) near Willow Street, now restored as a museum, reflects the architecture of the county's early German settlers. Stylistic characteristics of early Germanic houses illustrated in the Herr House include a steeply pitched roof, central chimney, and small casement windows. Traditional floorplans of Germanic houses typically included a *Küche* (kitchen), *Stube* (stoveroom, parlor/meeting room), *Kammer* (bed chamber), and often an *Arbeitsraum* (workroom). Perhaps the most recognized arrangement of spaces in Lancaster County Germanic style houses is the *Ernhaus* plan, also referred to as the *Flurküchen* plan,

Christian Herr House (1719)

Lancaster's most notable Germanic dwelling is located in West Lampeter Township. • John P. Herr

Johannes Hess Farmhouse (circa 1744)

The Hess Farmhouse in Warwick Township is a Germanic corner-post log house. • David B. Schneider

which consisted of three, or in the case of the Herr House, four rooms: a Küche with a large central "walk-in" fireplace forming approximately one-half of the house, a Stube and a Kammer forming the other half, and an occasional fourth room partitioned from the Küche, as found in the Herr House.[2]

Other room arrangements have been documented in Lancaster County. Among them is the *Durchgängigen Haus* plan, as illustrated by the Johannes Hess House (circa 1744) in Warwick Township, with a central hallway flanked by a Küche, and Arbeitsraum, Stube, and Kammer. Other early Germanic farmhouses include the Abraham Herr House (circa 1725) in Lancaster Township; the Theodorus Eby House (circa 1735) in Earl Township; the Martin Weybrecht House (circa 1741) in Manheim Township; the John Musser House (circa 1744) in Manor Township; the Benedict Eshleman House (circa 1759) in Conestoga Township; the Isaac Long Farmhouse (circa 1760) in Manheim Township; and the Christian Stauffer House (circa 1769, National Register) in East Lampeter Township. The Frederick White House in Paradise Township, while essentially Germanic in style, retains several distinctive characteristics reported to be of Huguenot influence, notably its double pegged window casing joinery detail.

Theodorus Eby Farmhouse (circa 1735)

This Germanic farmhouse in Earl Township was remodelled in the Federal style after 1800. • David B. Schneider

English

English style farmhouses differed stylistically from their German counterparts in that they were typically long and narrow, often were only one room deep, had low-pitched gable roofs with gable end chimneys, and contained double-hung sash windows. The William Downing House in Bart Township, built for a Quaker miller in 1747, is an example of this style. The Stubbs-Cutler House (circa 1767) in Drumore Township is another English style farmhouse. The David Davis House in Earl Township is

Reiber-Herr-Hershey House (circa 1775)

This house in Manor Township is a rather late example of a Germanic farmhouse. • David B. Schneider

William Downing Farmhouse (1747)

Traditional English farmhouse architecture is represented in this house in Bart Township. • John P. Herr

representative of a stylistic blend of features common to both German and English forms, having been constructed for a Welsh farmer and soon thereafter converted by Pennsylvania Germans.

During the mid to late eighteenth century, the distinctions between Germanic and traditional English styles diminished as aspects of each were accepted by the other. Farmhouses in rural Lancaster County built during the mid to late eighteenth century tended to be larger than their predecessors, generally reflecting the increased prosperity of Lancaster County agriculture. Typically, surviving farmhouses constructed during the period are two and one half stories tall, four to five bays wide, two bays deep, and relatively conservative stylistically.

Georgian

The Georgian style began to have influence in Lancaster County by about 1760 and lasted until the transition to the Federal style at the end of the century. Based on high style English architecture of the period, common elements of Georgian farmhouses in Lancaster County include both interior and exterior symmetry, elevations of between three and five bays width, and heights of two to three stories. Typical Georgian exterior ornamentation and features are found, such as flat arches over doors and windows, belt courses between upper floor levels, water tables, and pedimented doorways. The typical interior plan of local Georgian style houses includes four rooms per floor with a central stair hall.

While high style examples are found, such as Rock Ford, the home of Revolutionary War General Edward Hand (circa 1792, National Register), most Georgian houses in Lancaster County are vernacular interpretations of the style that often incorporate other stylistic influences. Most notable among these are the introduction of Germanic details, resulting from the adoption of the Georgian style by farmers of German lineage. Among the common examples of this style are the use of datestones, Germanic style bakeovens, and the presence of various Germanic decorative arts. Another variation is the so-called "two-thirds" Georgian style of the Quaker community that, although Georgian in character, lacks its traditional symmetry.

Additional examples of the Georgian style in Lancaster include: the John Jenkins Farmhouse (circa 1775) in Caernarvon Township; the Michael Baughman Farmhouse (circa 1790) in East Hempfield Township; the Jacob John Mohler Farmhouse (circa 1794) in Ephrata Township; the Abraham Landis Farmhouse (circa 1763) in East Lampeter Township; and the Andreas Graeff Farmhouse (circa 1767) in Lancaster Township.

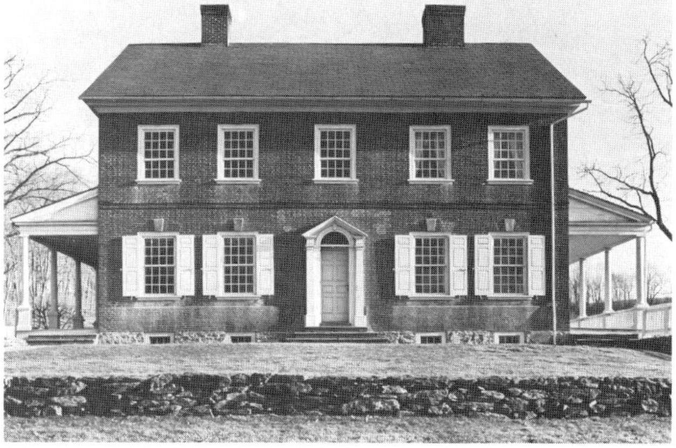

Rock Ford (circa 1790)

Note the symmetrical facade and pedimented entrance of this Georgian house in Lancaster City. • *John P. Herr*

Philip Friedrich Farmhouse (circa 1797)

This Georgian style farmhouse is located in Warwick Township. • *David B. Schneider*

John Pfautz Farmhouse (1813)

The Federal style in Lancaster County is well illustrated by this farmhouse in Warwick Township. *• Ed Worteck*

A detail of the cornice is shown at right. *• David B. Schneider*

A number of farmhouses represent the stylistic transition between the Federal and Georgian styles in Lancaster County in the late 1700s. Examples include the Philip Friedrich Farmhouse (circa 1797) in Warwick Township, which has a classic Georgian exterior with Federal interior details, and the John Haldeman Mansion (pre-1798, also called "Locust Grove," National Register) in Conoy Township.

Federal

By the late 1790s, the Federal style became popular in Lancaster County. It continued to be popular through the first third of the 1800s. The period immediately following the creation of the United States saw building designers seek to develop a national style that, while based on English neoclassical precedents, nevertheless reflected our indigenous materials, the importance of the nation's commitment to republican political and social institutions, and a need to exercise a certain economy in construction and ornamentation. The resulting Federal style was well suited to Lancaster County farmers who continued to use local materials and whose cultural patterns often found architectural expression in simplicity of construction and a rather restrained use of ornamentation. Many Lancaster County farmhouses retained the basic Georgian house form, including the four room plan, central hallway, and symmetrical facade. To this form, Federal style details were added, sometimes sophisticated but most often simplified. Cross-cultural influences that had been common during the Georgian period continued.

President James Buchanan's farmhouse at Wheatland (circa 1828, National Historic Landmark) is a recognized local example of the Federal style. Other Lancaster County examples include: the Peter Elser House (circa 1807) in Clay Township; the Nissley Farmhouse (circa 1811) in Rapho Township; the Henry Hertzler House (circa 1813) in Rapho Township; the John Pfautz Farmhouse (1813) in Warwick Township; the John Keller Farmhouse (circa 1813) in Ephrata Township; the Jacob Kirk mansion (circa 1815, National Register) in Little Britain Township; and the Jacob Keller Farmhouse (circa 1820) in Ephrata Township.

Pennsylvania Style

By the first third of the nineteenth century, a vernacular style, referred to here as the "Pennsylvania style," developed as a recognizable form throughout Lancaster County and the southeastern Pennsylvania region. Rooted in the symmetry of the Georgian style, the earliest Pennsylvania farmhouses are vernacular interpretations of the carry-over of its four room plan, central stairhall, and symmetrical facade into the Federal period. As the Pennsylvania style developed and flourished throughout the nineteenth and early twentieth centuries, it became, in essence, a vernacular platform upon which details from other architectural styles, the Federal, then the Greek Revival, and later, the Victorian, were applied. Pennsylvania style farmhouses are therefore identifiable more by massing, fenestration, and floorplan than by stylistic features. Among the traits common to all examples of the style are a rectangular, often square or almost square, core shape; a lateral gable roof over the core; two and one half stories; and a symmetrical facade of three, four, or five bays in width. The four room plan of the Georgian and Federal periods is typically continued, although stair location varies widely, and the stairhall is often deleted in favor of a simple closed stair.

The most common materials are stone or brick, with frame and log examples also found. Other features typical of the style are the use of datestones, often listing the original owners of the house, and bell cupolas, in which a farm bell was hung. Many examples have original or later wings, typically placed as a rear ell, a recessed side wing with a front porch, or a rear

Michael Schindel Farmhouse (1816)

The Michael Schindel Farmhouse in Manor Township illustrates the development of the Pennsylvania style as a distinct vernacular form. While it is also representative of a rural Federal style house, the simplicity of its detail contrasts with the John Pfautz Farmhouse on the previous page. Note also the presence of the recessed side wing (right) on this five-bay example of a Pennsylvania style farmhouse with a central entrance. • *Ed Worteck*

John Herr Farmhouse (1813)

The use of vernacular Federal style elements in a Pennsylvania style farmhouse is illustrated by this detail of the entrance surround and transom. • *David B. Schneider*

Jacob Siegrist Farmhouse (1822)

This is a four-bay example of a Pennsylvania style farmhouse with an off-center entrance. Note the rectangular shape and two room depth. • *David B. Schneider*

tee. Entrance placement varies extensively, although the majority of documented four-bay Lancaster County examples have paired central entrances, and the majority of five-bay examples have single central entrances.

Because the Pennsylvania style began as a vernacular form of the Federal style, early examples can overlap both styles. The John Pfautz House in Warwick Township is a Federal style house that shares the basic characteristics of the Pennsylvania style. The more vernacular character of the Michael Schindel House (1816) in Manor Township illustrates the divergence of the Pennsylvania style, with its simplicity of detail and the addition of a recessed side wing. Other examples of early Pennsylvania style farmhouses exhibiting Federal style details are the John Weidler Farmhouse (1813), in Manheim Township; the John Herr Farmhouse (1813), in Manor Township; and the Jacob Siegrist Farmhouse (1816), also in Manor Township.

As the Federal style waned in popularity, both nationally and locally, the Pennsylvania style became more recognizable as a distinct form. The basic rectangular core, with its two

Andrew Brubaker Farmhouse (1839)

This five-bay Pennsylvania style farmhouse in East Hempfield Township has a rear ell wing with a bell cupola and a rear kitchen house. • *David B. Schneider*

Andrew Brubaker Farmhouse (1839)

Note the window lintel with corner blocks and the paired datestones. The right stone records the craftsmen who built the house. • *David B. Schneider*

William & Mary Smith Farmhouse (1853)

The house was a classic example of a four-bay Pennsylvania style farmhouse • David B. Schneider

William & Mary Smith Farmhouse (1853)

Many Pennsylvania style farmhouses have rear tee or ell wings. Note that this wing slopes from three stories to one story. • David B. Schneider

Jacob Kurtz Farmhouse (1861)

This four-bay Pennsylvania style house had paired entrances and a recessed side wing. • David B. Schneider

and one-half story height, symmetrical facade, and four room plan, remained constant. Details of individual farmhouses were often influenced by the architectural styles popular at the time of their construction, and the level of ornamentation varied greatly. While the most recognizable Pennsylvania style farmhouses tend to be simply detailed, it is also common to find examples with elaborate Greek Revival or Victorian elements.

The William and Mary Smith Farmhouse (1853) in East Hempfield Township is typical of the style. A sampling of other examples includes the Shreiner Farmhouse (circa 1828) in Manheim Township; the Joseph Good Farmhouse (circa 1836) in Pequea Township; and the John and Susan Eshleman Farmhouse (1851) and the Habecker Farmhouse (1878) in East Hempfield Township. At least one six-bay example of the style has been documented in the Charles Bamford Farmhouse (circa 1860) in East Hempfield Township, which also exhibits both Greek Revival and Victorian stylistic elements. An interesting example of a late nineteenth-century conversion of a Germanic style farmhouse into a Pennsylvania style farmhouse has been documented in the Peter and Catharina Reyer Farmhouse (1792, circa 1880) in Ephrata Township.

The "Farms in Berks County, PA" multiple property nomination to the National Register of Historic Places identifies these common farmhouses as "Pennsylvania German" style buildings because of their retention of basic elements of Pennsylvania German craftsmanship.[3] Yet farmhouses of this type were built by non-Pennsylvania Germans, and the style is rooted in both the Georgian and Federal styles. Other historians, notable Henry Glassie and Allen Noble, have referred to these houses as simply "four over four" in reference to their basic floorplan, but such a description is also applicable to both the Georgian and Federal styles. Joseph Glass, who identified a number of the characteristics of these farmhouses in *The Pennsylvania Culture Region*, all of which are well represented in Lancaster County, does not assign a specific name to the style.[4] In *Architecture of the Pennsylvania Dutch Country, 1700-1900*, Henry Kauffman includes several Pennsylvania style examples in his "Post Georgian style," but at least three of the houses in his subsequent chapter on the Victorian style are similar to the Pennsylvania style.[5] The choice of the term "Pennsylvania style" here therefore reflects the origin of the style as a vernacular tradition that developed as a distinct form in southeastern Pennsylvania and spread to other regions, in much the same manner as did the more widely recognized Pennsylvania barn.

Greek Revival

No farming-related examples of high style Greek Revival architecture have been identified in the Lancaster County Historic Sites Survey. The Leaman Mansion in Paradise Township, the county's primary example of the columned Greek Revival style, was built for a doctor and was later acquired by a gentleman farmer. While pure examples of the Greek Revival style are rare, vernacular Pennsylvania style houses with Greek Revival details continued to be popular throughout the the nineteenth century. For example, the Christian O. Herr House (1860) in Manor Township retained a particularly noteworthy Greek Revival style entrance until it was demolished in 1992.

Victorian

The broad range of Victorian styles became popular in Lancaster County in the late nineteenth century. Second Empire, Italianate, Gothic Revival, and Queen Anne style farmhouses are found, as are eclectic examples mixing stylistic elements. Pennsylvania style farmhouses utilizing ornamentation and features from each of the Victorian styles are also common. As in past stylistic periods, ornamentation in most Lancaster County farmhouses is most often quite restrained, although high style examples are occasionally found.

The Second Empire style, typified by the use of a Mansard roof, found acceptance locally primarily from circa 1860 to 1900 and is represented by houses such as the Burkholder-Rush Farmhouse (circa 1875) in Pequea Township, and the Watts Mansion (circa 1873-1874) in East Donegal Township. The David Mayer

Christian O. Herr Farmhouse (1860)

The Greek Revival style often found expression in rural Pennsylvania style farmhouses. • John P. Herr

Charles Bamford Farmhouse (circa 1860)

This rare six-bay Pennsylvania style farmhouse in East Hempfield Township combines Greek Revival and Victorian details. • David B. Schneider

L. R. Reist Farmhouse (1875)

Lancaster retains a wealth of Victorian farmhouses such as this Manheim Township example. • David B. Schneider

David Mayer Farmhouse (circa 1870 - 1874)

The David Mayer Farmhouse in Manheim Township is an example of the Italianate style and is one of Lancaster's landmark rural residences. *•Ed Worteck*

Silas K. Eshleman Farmhouse (circa 1875)

This house in Paradise Township illustrates both elements of the Victorian Gothic and the Italianate styles. The blending of stylistic details was common during the Victorian period. *• Ed Worteck*

Farmhouse (circa 1870) in Manheim Township is a particularly good example of the Italianate style, which began to become popular in the 1850s and continued to be used to the turn of the century.

The Gothic Revival style was popular in Lancaster County from the 1840s through the 1870s. The Silas Eshelman Farmhouse (circa 1870) in Paradise Township is a frame Gothic Revival style dwelling with Italianate elements. The Mellinger Farmhouse (1893) in Manor Township is a good example of the Queen Anne style.

Often details and elements from a variety of Victorian styles are found within a single farmhouse. The John Kendig Mansion (circa 1874-1886) in West Lampeter Township is such a high Victorian eclectic design. Among the Pennsylvania style farmhouses exhibiting Victorian details are the Christian Herr Farmhouse (1864) in East Lampeter Township, the Abraham Brubaker Farmhouse (1876) in East Hempfield Township, and the Daniel and Margreda Weldy Farmhouse (1901) in West Hempfield Township.

Twentieth Century

Early twentieth-century architectural styles are also represented in Lancaster County farmhouses, although somewhat infrequently. Variations of the American Foursquare and other simplified Victorian derivatives are found, as are Craftsman and Bungalow style houses. In addition, there are also a number of early twentieth-century houses constructed of patterned concrete block. An example of an early twentieth-century Lancaster farmhouse is the Shenk Farmhouse (1901) in Manor Township. The John Brenneman Farm (1900) in Pequea Township is representative of a simple early twentieth-century patterned block farmhouse.

Esbenshade Farmhouse (Early 20th Century)

The Esbenshade Farmhouse was reflective of the Pennsylvania style in its general massing and symmetrical five-bay facade. Note its Colonial Revival details, such as the tripartite attic window, its cornice, and its pedimented entry porch. *• David B. Schneider*

The Barn

The Pennsylvania Barn, by Robert F. Ensminger

The paragraphs below are reprinted and edited from the "Farms in Berks County, PA" multiple property nomination to the National Register of Historic Places by permission of Mr. Ensminger and the Berks County Conservancy.

In the early eighteenth century, permanent barn structures were built in Lancaster County by Germanic pioneers. These early types were modeled after similar structures from their European homelands. The earliest barns were small and simple and consisted of a single log crib plus various attached sheds. They could house some cattle, feed, grains and a few hand implements.

This standard ground barn plan soon replaced the small first generation pioneer barns. This barn had two log cribs separated by a central threshing floor combined under a gable roof. The ground level of each log crib provided stabling space while the space above the stables and beneath the roof provided storage for hay and straw. This multiple purpose ground level barn, or Grundscheier, was modeled after similar structures common in the German Palatinate whose masonry and half-timbering were the common materials of construction. The almost universal use of log construction was most appropriate on the Pennsylvania frontier where the precedent had already been effectively established by earlier Swedish settlers in the lower Delaware Valley who passed on the technology to Germanic pioneers who moved farther inland.

Germanic settlers from eastern Switzerland introduced the log, two-level, banked, forebay barn during the same period when the Grundscheier was introduced. The Pennsylvania version closely resembled its Swiss prototype from Pratigau in Canton Graubun-

The Eck Barn (1804)

Located in Manheim Township, the Eck Barn is an example of a Pennsylvania barn with later frame additions. Note the steep pitch of the roof. *• Ed Worteck*

Jacob Shank Barn (1835)

The interior of the Jacob Shank Barn was typical of many Pennsylvania barns. Note the timber framing and the ventilation slits in the gable wall. • Robert Lowing

den. The upper level had two log crib mows and central threshing floor accessed by a ramp or bank on the back side. A forebay extended from this level six to eight feet beyond the lower stable front wall providing protection for the stable doors below. Commonly called the "Sweitzer" barn, this structure was characterized by an unsupported, cantilevered forebay which also provided a diagnostic asymmetrical gable end configuration. Both the Grundscheier and Sweitzer barn were commonly built in all Germanic regions of southeastern Pennsylvania.

During the latter part of the eighteenth century, commercial farming gradually replaced subsistence agriculture in southeastern Pennsylvania. Regional commercial market towns, such as Lancaster and Reading, and the development of an interconnecting road net focused on Philadelphia, stimulated the expansion of commercial agriculture and the evolution of larger barns. Sweitzer barns displaced the smaller Grudscheier and were built much larger and utilized stone construction. These classical Sweitzer barns came to dominate the early nineteenth-century landscape.

During the golden age of Pennsylvania agriculture from 1790 to 1840, the rapid expansion of commercial agriculture stimulated barn evolution resulting in a variety of styles and sizes of forebay barns which had come to characterize and symbolize the rural Pennsylvania Germanic landscape of Lancaster and surrounding counties. The name "Pennsylvania barn" has come to be used exclusively to designate the forebay bank barns which developed here.

The first major barn style change resulted in what became the most abundant class of Pennsylvania barns in southeastern Pennsylvania and Lancaster County. It occurred when the asymmetrical cantilevered forebay was eliminated. The forebay form, however, was retained by recessing the lower front stable wall four to six feet back under the upper level, thus recreating the forebay overhang which was now part of the main barn frame. This resulted in a symmetrical gable wall configuration and is named the standard Pennsylvania barn.

Various versions of this form were built throughout the nineteenth century. The earliest ones were stone construction with lower gable end walls completely closing the recessed forebay. Many of these utilized "L"-shaped pillars, or Peilers, to strengthen and support the frame forebay front wall producing an L-shaped alcove, or Peiler Eck, on either side of the front stable wall. Most standard barns built after 1850 were frame with timber frame bent construction. Many of these eliminated the extended foundation end-wall support producing an "open forebay" cantile-

Hershey Barn (19th Century)

The Hershey Barn in East Hempfield Township is a typical frame Pennsylvania barn. Note the earthen ramp leading to the barn doors. The frame tobacco shed (left) and crib (right) are later additions. • *Ed Worteck*

Brackbill Barn (1795)

Built by a German farmer, the Brackbill Barn in Paradise Township has a somewhat English character in its lack of a forebay and its arched entrances. The cross-shaped threshing level is arranged so that hay mows at each corner rise from the ground level to the rafters. • *David B. Schneider*

John Eby Barn (1860)

The John Eby Barn in Manheim Township is a frame Pennsylvania style barn. Note the elaborately detailed ventilators at the upper portion of the forebay. *• Ed Worteck*

Keneagy Barn (19th Century)

The Keneagy Barn in Paradise Township is one of only a handful of Lancaster County examples of a triple deck barn. Note the bridge at left, entering a third floor level near the top of the barn. Access to the second level is through the doors at right. A forebay is located on the opposite side. • David B. Schneider

vered configuration. Later examples frequently included posts below the forebay sill for additional support.

Another evolutionary development which originated in adjacent Chester County and spread north and west was the addition of a large, extended forebay straw shed to various existing barns. The resulting deep 20 to 25 foot forebays necessarily required additional support provided by posts. In Chester County, conical stone columns of English origin were frequently used.

An alternative strategy for barn enlargement involved the inclusion of rear outsheds on either side of the barn bank. This plan, which emerged in Lancaster County in the early nineteenth century, soon diffused west and north. The outshed appendage, which included the basement stable, usually provided a granary function adjacent to the upper level mow and threshing bays.

In the nineteenth century, many barns, were amended by enclosing the entire rear bank wall creating a large "rampshed" storage area. Other barns were enlarged by the addition of a large shed projecting from the forebay producing an "L"-shaped wing with an additional gable wall. Both additions permitted storage of large amounts of straw and hay.

Although the building and rebuilding of two-level Pennsylvania barns continued into the twentieth century, most new barns built after 1900 were specialized, dairy barns designed and promoted by agricultural colleges and journals. A fascinating example is provided by the barn at Grandview Farm near Wernersville, Berks County, now owned and operated by Adam Schaeffer. In 1901, an existing 1849 stone, standard Pennsylvania barn was greatly enlarged by extending the barn lengthwise for an additional 100 feet. This two-level banked addition replicated the earlier timber frame bents creating two additional mows and threshing floors. The under forebay space of the addition was enclosed, thus enlarging the stable area providing room for 60 cows in two longitudinal rows of wooden stalls each served by a water line and drinking bowl. The stable was ventilated by wooden ducts which removed stale air to a row of gable dormers along each of the eaves and to large cupolas along the roof ridge. Large hay drop holes enclosed in two sheds were built onto the barn's bank side. One of the sheds also enclosed the silo as an integral part of the barn's design. This was, indeed, high technology for that period, and it continues to function well today with no revision.

Barn evolution in the eighteenth, nineteenth and twentieth centuries has produced a varied and interesting assemblage of types. In Lancaster County, the traditional Pennsylvania barn with its forebay and bank, has maintained its dominance of the landscape revealing the strength of Swiss and Germanic traditions.

English Barns

Barns constructed by the English, Welsh, and Scotch-Irish in the eighteenth through the early nineteenth century developed differently from the Pennsylvania barn. Typically rectangular with gable roofs, they were constructed on flat sites rather than banked ones and had no projecting forebay. Frame and stone were common materials, although early log examples undoubtedly were present. Many of the surviving examples identified in the Lancaster County Historic Sites Survey are located within the northwestern section of the county. They are generally constructed of stone and typically have seen a great deal of alteration in response to changing farm practices. Among the examples identified is a barn located adjacent to Spring Grove Forge in East Earl Township.

Kirk-Haines Barn (19th Century)

This barn in Little Britain Township has an unusually tall supported forebay • Heidi M. Pawlowski

Christian Herr Farm, Tobacco Barn (19th Century)

Typical of tobacco barns found throughout Lancaster County, the tobacco barn on the Christian Herr Farm in East Lampeter Township has tall vertical wall ventilators and roof ventilators to dry the tobacco hung inside. • *Ed Worteck*

Plain Sect Barns

Amish, Mennonite, and other plain sect groups within Lancaster County historically built and utilized Pennsylvania barns. With increasing reliance on milk and cattle production throughout the nineteenth century, forebays were commonly enclosed, and straw and cattle sheds attached. The appearance of many Pennsylvania barns operated by plain sect farmers today have become almost unrecognizable due to these additions. Yet this practice is a continuance of the traditional Lancaster County response to changing markets and technologies.

Tobacco Barns

In Lancaster County, tobacco barns became an integral component of the farm by the mid nineteenth century. Typically constructed of wood, these large open storage facilities usually have a system of vertical or horizontal ventilators within their sidewalls and often roof ventilators. Tobacco barns are generally constructed on a raised stone foundation to provide for a tobacco cellar, or dampening room, and stripping room. Within the barn, a series of poles and racks provide a framework from which tobacco is hung to cure. The majority of operating farm complexes identified in the Lancaster County Historic Sites Survey possess one, if not more, tobacco barns.

Domestic Outbuildings

Springhouses

Springhouses, from which water was provided for domestic uses, were almost universal elements of the Lancaster County farm. The purpose of the springhouse was to shelter the water supply. They generally consisted of small rectangular buildings, often recessed into the ground, of stone, brick, or wood. Some examples were more than one story in height and provided a living or working space on the upper level with a springroom below. Storage was often provided for perishables, milk, and butter in the springroom and occasionally spaces were provided for laundering, cooking apple butter, making soap, or even butchering. Springhouses continue to be an integral part of Lancaster County's agricultural landscape, and many farms identified in the Historic Sites Survey retain them.

David Davis Springhouse

This stone springhouse is located in Earl Township.
• David B. Schneider

Kitchens

Some form of building detached from the farmhouse, in which butchering, the processing of dairy products, and other domestic activities were carried out appears to have been common to Lancaster County farms from the period of settlement. Often springhouses served this function, although early references to out kitchens or back kitchens can be found. By the latter part of the eighteenth century, the food preparation functions of springhouses and/or out kitchens were typically carried out in what became known as a summer kitchen or summer house. Cooking and preserving of foods were done in these structures during warm weather in order to keep the heat out of the main dwelling. In addition, summer kitchens often served as a washhouse or butcherhouse. Summer kitchens were located in close proximity to the farmhouse and often attached to it by an open breezeway. During the mid nineteenth century, summer kitchen functions were often relocated to basement rooms constructed for this purpose. The Michael Schindel House in Manor Township is among many sites identified in the Lancaster County Historic Sites Survey that retain excellent examples of summer kitchens.

Michael Schindel Kitchen

Located in Manor Township, this frame summer kitchen is typical of many in the county. *• David B. Schneider*

Michael Schindel Smokehouse

• David B. Schneider

Bakeovens

Among the outbuildings devoted to the processing of foods for domestic consumption were the bakeoven, smokehouse, and butcherhouse. The bakeoven was used for baking breads and pies, roasting cornmeal and nuts, and drying vegetables. Bakeovens were commonly constructed as free standing exterior structures, such as the example at the Caushey-Ankrim House in Drumore Township (circa 1765) or attached to the rear of interior cooking fireplaces at the farmhouse or summer kitchen, as is the case in the summer kitchen of the Shreiner Farm in Manheim Township.

Smokehouses

Smokehouses, in which meats were cured, were and are also common. Typically detached buildings of stone or brick, examples include the brick smokehouse at the Schindel House in Manor Township and the brick and stone smokehouse at the Peter LeFever House in Strasburg Township. It was also common to find smoking rooms or chambers adjacent to the chimney in the attics of farmhouses, as in the Martin Barr House in Strasburg Township.

Butcherhouses

Butchering of meats was often accomplished in a butcherhouse, where pigs and beef could be slaughtered, hung, cleaned, cut, and processed. Stoves with large kettles to provide hot water to render the lard out of the carcasses were common. In addition to butcherhouses, this activity was often accomplished in the washhouse or in the basement of the farmhouse.

Washhouses

The other prominent domestic outbuilding was the washhouse in which clothes were cleaned. The washhouse contained a fireplace or stove to heat water and shelves or benches for tubs and washing equipment. Washhouses continue to be an active component of many of Lancaster County's plain sect farms.

Cave or Ground Cellars

Before the days of refrigeration, nearly every farmstead had a cave or ground cellar for the storage and preservation of food. Both detached cellars, such as the example found at the Circle Creek Farm in East Donegal Township, and those beneath or adjacent to the house were common in Lancaster County. They were typically constructed of stone, often whitewashed or parged on the interior, and had arched ceilings. Floors were typically earth, brick, or stone. The cellar provided cool storage area for fruit, vegetables, and dairy products.

Ice Houses

Prior to the invention of electric refrigeration in the twentieth century, the cutting and storage of ice was an important seasonal event on the farm during the winter months. Ice was cut from streams or ponds and stored in an icehouse for use during the remainder of the year. Icehouses were typically constructed of stone

Ice House at Rohrer's Mill

The farm complex at Rohrer's Mill in Paradise Township retains this nineteenth-century frame ice house as well as its associated ponds. *• Ed Worteck*

Silas K. Eshleman Farm, Privy

The architecture of the privy mirrors the adjacent farmhouse and barns. *• Ed Worteck*

or wood and often had double insulated or cork-lined walls. The Rohrer Farm and Mill in Paradise Township retains two icehouses and a series of ponds and races from which ice was collected.

Privies

Until the twentieth century, all farms had outdoor privies. Construction materials and sizes varied, as did architectural embellishment. Simple examples, such as the privy located on the Benjamin Herr Farm in East Lampeter Township, and more elaborate ones, such as the outhouse on the Silas K. Eshelman Farm in Paradise Township that mimics the Victorian Gothic architecture of its adjoining farm buildings, remain throughout the county. The outdoor privy continues to be used today within the county's plain sect community, principally at rural schools.

Farm Outbuildings

Silos

Silos were introduced on Lancaster County farms beginning in the last third of the nineteenth century, yet did not become commonplace until about 1920. They are typically round; constructed of wood, concrete, tile, or metal; and located adjacent to the barn. The visual character of Lancaster County has evolved through the introduction of this vertically prominent element within the landscape. Much as the historic character of the skyline of Lancaster City was defined by church steeples. Similarly, the skyline of rural Lancaster County is now shaped by silos.

Milkhouses

In the earliest farms, milk was typically cooled in the waters of the springhouse, ground cellar, or basement. By the late eighteenth century, specialized milkhouses, typically built of stone and near or adjacent to the barn, became commonplace. By the late nineteenth century, milkhouses constructed of wood, concrete block, brick, or tile were introduced. Increasingly strict regulations for the handling of milk have resulted in further evolution of milkhouses. In Old Order Amish farms and other plain sects that do not otherwise utilize electricity, modern refrigeration is found in the milkhouse.

Other Outbuildings

Among the other common outbuildings historically found on Lancaster County farms are malthouses, stillhouses, and wineries, in which beverages were produced, principally for domestic consumption; corn cribs, both detached and as appendages to barns; tool sheds; pig pens, sheepfolds, and chickenhouses; and, typically in larger farms, hay barns, which were used to store excess hay that could not otherwise be accommodated within the main barn.

Unidentified Farm, Silos

Silos became commonplace in Lancaster County in the early twentieth century. *• Ed Worteck*

Christian Habecker Farm, Milkhouse

This small frame milkhouse is located on the Christian Habecker Farm in East Hempfield Township.
• Ed Worteck

Other Site Features

Garden, Yard, and Orchard

Like the farm buildings themselves, the land was divided among largely domestic and farming activities. The domestic spaces adjacent to the farmland were and are typically used for vegetable and flower gardens. Vegetable gardens and gardens for the production of cutting flowers are typically located to the rear or side of the farmhouse. Often planting beds of flowers were sited to the front of the house and along walks leading to it. Fruit and shade trees are also commonly found around the house. Historically, gardens were often surrounded by picket fences or even by stone walls. Orchards, when used primarily for domestic use, were often located in close proximity to the farmhouse.

Cropland

In the eighteenth century, Pennsylvania farmers developed a pattern of farming that was far different than what had been generally practiced in Europe. Unlike the southern plantation system, where much of the labor was provided by servants, or the New England practice of cultivating gardens, plentiful land, rich soils, and good farming practices, allowed each farmer to develop a series of his own "fields." Croplands, like the farm buildings themselves, were and are constantly adapted to new farming technologies and market conditions. With the acceptance of technological advances by Pennsylvania farmers, such as the use of horse drawn equipment and, later, mass produced farm machinery, fields expanded dramatically in size. In more recent years, contour farming

Cropland

Historic farm buildings often exist within the distinctive visual context of wide expanses of cropland. *• Ed Worteck*

and soil conservation practices have continued this evolutionary process.

Meadow, Stream, and Woodlot

Meadows were typically used for cattle, and often winter wheat was grown on them. They generally were created in the lowlands surrounding streams. Permanent pastures developed after about 1750. Streams and springs are important features, as they provide fresh water for both the farm family and its animals. In Lancaster County, much of the thick forest that covered the land when the settlers arrived disappeared by the late 1800s. Small woodlots were retained, generally where topography made farming difficult.

Fencing

Fencing is a common feature of the Lancaster County farm that has changed considerably over time. Early settlers utilized the products of the forests they cleared to construct wooden fences, such as the worm fence, or snake fence, constructed of split rails built in a zig-zag course that did not require posts. The post and rail fence was often used as a replacement for the worm fence. Although it required more time and labor to erect, it was more substantial and took up less space. Posts of chestnut or locust were preferred, as they were long lasting.

As farms became more permanently established, stone walls and rail fences became commonplace. Stone walls typically separated the barnyard, served as retaining walls where slop-

Farm Scene

Along with cropland, meadows, streams, and woodlots are common elements of the farming landscape. • *Ed Worteck*

ing sites surround the domestic buildings, and surrounded family burial plots. Otherwise, stone boundary walls were either not common to Lancaster County or have simply not survived. Fences served to keep farm animals within the farm, wild animals out, divided fields, and established boundary lines. In the county's early history, rail fencing or vegetation such as hedgerows or tree lines provided boundaries for farms and fields.

In the late nineteenth century, wire fencing was introduced. Iron and cast iron fencing became increasingly popular for house yards and family burial grounds in the mid to late nineteenth century. Of more recent origin is the board fence, often used for horses and cattle pastures. Fences around the farmhouse were more ornamental in character, favorites being the wrought iron fence and the pale or picket fence. Particularly in eastern Lancaster County, fencing as boundary or field demarcation has largely disappeared. Farmers generally use natural landmarks, such as streams, scattered trees and other natural features, or remnants of earlier demarcation of land use, such as ditches, fence posts, etc., to delineate their lands.

Lime Kilns

Lime, used to improve the productivity of the soil and as a building material, was an important aspect of farming for Lancaster's Pennsylvania German community. By the 1750s, lime kilns, used to extract lime from limestone, were common throughout the county. The operation of lime kilns was often cooperative, with several farms sharing a single kiln. The kilns themselves varied in design and construction. The most common examples remaining today are constructed of stone, have arched or pointed openings, and are located on steep slopes or within woodlots. Made technologically obsolete by the introduction of commercially available lime in the mid nineteenth century, and its popular acceptance by the late nineteenth to early twentieth century, most surviving kilns have fallen into disrepair. A typical example of lime kilns identified in the survey is found on the Martin Barr Farm in Strasburg Township.

Buch Family Burial Ground

• David B. Schneider

Family Burial Grounds

The practice of establishing small family burial plots was common on Lancaster County farms throughout the eighteenth and nineteenth centuries. Often surrounded by low stone walls or, in later examples, iron fencing, these plots were typically small in size and sited away from the farmstead, often overlooking it. Surviving gravestones and markers are typically plain or iconographic in style, although some monumental elements are occasionally found. Although most family graveyards retain no more than one or two dozen markers, and often fewer, many contain unmarked graves of infants or graves for which markers have disappeared due to deterioration or removal. Examples include the Baughman Family Burial Ground (circa 1774) near Georgetown in Bart Township and the Shirk Family Burial Ground (circa 1820) in Manheim Township.

For European-Americans, the family burial ground was a phenomenon for which their was little precedent in their homelands. While the reasons they developed in this country are not well documented, they survive as an important aspect of the eighteenth and nineteenth-century agricultural landscape. The frequency with which these burial grounds occurred historically is difficult to estimate, as many were too small to be shown on county atlases. Families often did not specifically reserve their plots on their deeds, and many have disappeared as a result of vandalism or deterioration.

Plain Sect Farms

As much as farming has given Lancaster County a national identity, so too have the plain religious sects that settled here in the eighteenth century and remain here today. Not only do these groups preserve many of the oldest working farms in the county, they often do so using a unique mix of modern and pre-industrial technologies. The character of the county as a whole is strengthened and diversified by the continued experience gained from being able to witness horse drawn agricultural machinery and buggies. The character of the community is even affected at nighttime, when vast stretches of farmland are pitch dark because of the absence of electrical lighting. These visual aspects combine with significant cultural traditions to form one of the county's most unique dimensions.

These cultural patterns are reflected in the farms of the plain people. Simple in design and typically lacking ornament, the sites are nonetheless often complex groupings of farm buildings with numerous generations of additions. The telescoping visual effect of a farmhouse with successive Dawdy Haus additions and the sprawling evolution of the barns are examples of this.[5] The continued presence of windmills, water engines, and other such structures are characteristic.

There are two major types of plain sect farms in Lancaster County: those farms representing "pure" plain sect farms, where the

An Amish Farmstead

• Ed Worteck

An Amish Farmstead

• Ed Worteck

present character of the farm is wholly a product of its development by plain people, and "altered" plain sect farms, where farms of other cultural groups have been acquired by plain sect farmers and yet still retain aspects of their previous character. An example of the former is the Stoltzfus Farm in Leacock Township and of the latter is the Benjamin Brackbill Farm in Paradise Township.

Water and Wind Power

Relatively common features of many, particularly plain sect, farms in Lancaster County are systems to harness water and wind to produce power for farming operations. Small dams and races channel water to wheels used to turn pulleys, which are often arranged in complex systems. The pulleys in turn transmit mechanical power that is used for a variety of farming and domestic activities. Windmills are also used in much the same fashion.

Lost Agricultural Context

Since the early twentieth century, an increasing amount of Lancaster County farmland has given way to suburban development. While much of this merely expanded the boundaries of existing urban centers, suburban development has also been scattered throughout the rural areas of the county. More than 92,500 acres of farmland have been lost to development since 1959.

Because of the rapid growth and development experienced in Lancaster County in recent years, many historic farm buildings and farmsteads have lost the context of their farmland. Yet individually and by their pattern on the land they are valuable both for what they contribute to a broader understanding of the overwhelming historical significance of farming in Lancaster County and as reminders of the prior agricultural function of areas now in other land uses.

Silas K. Eshleman Barns

Like the privy pictured previously, the barns of the Silas K. Eshleman Farm reflect the Victorian architecture of the farmhouse. Yet they are also representative of the simple elegance of Lancaster County's farming buildings and a reminder that the commonplace elements of Lancaster's cultural landscape are what define its essential character. • *Ed Worteck*

4: Case Studies

Lancaster County is unique for its abundance of continuously operated historic farms. These farms remain as viable economic enterprises because they adapted to technological advances and changing market conditions. The county is fortunate that farm buildings dating from the eighteenth century through the twentieth century continue to be used for their original purposes. Yet the landscape is changing, and many farms have been compromised by commercial and residential development.

The sample of farms that follows illustrates only a small segment of the full diversity of Lancaster County's farming heritage. They represent various periods of the agricultural history of the county and the diversity of its cultural influences.

All of the sites chosen are typical of commonplace Lancaster County farms. The David Davis Farm, Christian Habecker Farm, Shreiner Farm, and Windom Mill Farm are examples of farms that illustrate the evolution of farming and the changing character of farm buildings. The Eastern Mill Creek Valley represents the broad pattern of Lancaster County's farming landscape.

With the increasing suburbanization of Lancaster County, many farming related buildings have lost the context of their associated farmland, and, in some cases, even their associated farm complexes. Yet, as in the case of the Bausman Farmstead, they remain as significant vestiges of the historic importance of farming in Lancaster County.

A Scene in the Eastern Mill Creek Valley

• Ed Worteck

David Davis Farm

Earl Township

The David Davis Farm is an excellent example of the evolutionary nature of many Lancaster County farms through several generations of ownership by differing cultural groups. While it evidences the eighteenth-century Welsh settlement in this portion of the county, it also exhibits the influences of subsequent generations of Amish and Mennonite farmers. The largely intact farmstead, consisting of a farmhouse, bank barn, springhouse, tobacco barns, sheds, and a family burial ground includes features dating from circa 1750 through the early twentieth century.

The David Davis Farm was patented by Thomas, John, and Richard Penn to David Davis, a Baptist from Cardiganshire, Wales, in 1744. Davis obtained a warrant and surveyed the 297 acre property by 1730. A shopkeeper who operated a store in Blue Ball, Davis lived on the farm until his death in 1773 at age 91. He is buried in the family burial ground on the farm. The property passed by David Davis' will, dated 1768, to his son John, who took full possession by a deed dated 19 October 1784.

The farm was purchased by Christian Summers for £2,500 on 11 April 1785. He sold one

The Farmhouse

Structural evidence suggests that at least a portion of the western stone section of the building was constructed circa 1750 for David Davis. The house was extensively remodelled in1787 by Christian Summers, an Amishman. After 1815, the two bay stone addition was constructed to the east by the Martin family. *• Ed Worteck*

The Barn

The barn is typical of the Classic Sweitzer type. Its steeply pitched roof is consistent with family tradition that holds it was built circa 1780. • *David B. Schneider*

half of the farm, 148.5 acres, to Jacob Sensenig shortly afterward, on 21 May 1785. A wooden dateboard survives (but has now been removed from the farmhouse) which reads: "1787 Christian Summers and Froney his wife. Renewed by John Martin." Stylistic and physical evidence suggests that Summers extensively remodeled an earlier house built by Davis. The dateboard appears to have been installed by Martin, who may have assumed that Summers built the house. Summers was the first of three Amish owners between 1785 and 1796. After his death in 1790, his estate sold the then 170 acre farm at auction. Jacob Hertzler purchased it on 3 September 1790 and transferred it to Jost Yoder on 23 September.

In 1796, the 170 acre farm was acquired for £1,750 by Martin Martin, a Swiss Mennonite pioneer who settled in Lancaster County and lived on an adjoining farm. Martin's son Christian purchased 104.5 acres of the farm in 1799 and sold it to his son John Martin in 1849. Subsequent owners have included Isaac Martin, Sadoc Martin, Isaac N. Martin, and Paul Martin, its present owner.

The David Davis Farm is located within the rich limestone plain that forms the central core of Lancaster County's most productive farmland. The farm is surrounded by agricultural land with the exception of limited development that has occurred along U. S. Route 322, immediately north of the site.

Family Burial Ground

David Davis and and his wife Hannah, as well as members of the Martin family, are buried in the small family burial ground. • *David B. Schneider*

Christian Habecker Farm

East Hempfield Township

The Christian Habecker Farm is a fine example of a working Lancaster County farm that has retained its architectural and agricultural significance in the midst of the suburban growth that surrounds it. Although the first permanent farmstead was constructed on the property in the late 1700s, the farm has been owned by the Habecker family since 1853, and all existing buildings date from their occupancy. The now ninety acre farm remains under cultivation.

The present farm is part of a 230 acre tract acquired by Abraham Brubaker in 1793, and on which he began to construct a permanent farmstead in 1795. Brubaker was the son of Hans Jacob Brubaker, thought to be a cousin to Hans Brubaker, who also settled in Lancaster County in 1717. After Abraham's death, the property passed to his daughter Elizabeth and her husband Christian Mellinger on 6 February 1817. In 1840, the Mellingers sold the property to Michael Seitz who, in turn, sold it as a 116 acre farm to Christian B. and Elizabeth K. Habecker on 15 March 1853. After Christian's death in 1858, the property remained in the possession of his wife and heirs. The present barn bears a datestone that reads: "REBUILT / BY THE HEIRS / OF / CHRIST B. HABECKER / A.D. 1869."

The farm was later acquired by Christian and Elizabeth's son Christian H. Habecker who married Susan Fry Shenk in 1876. According to its datestone, which serves as the sill to the front entrance, the present farmhouse was constructed for them later that year. By the time an 1899 county atlas was prepared, the property was listed as 109 acres under the ownership of Christian Habecker.

Christian Habecker died on 1 January 1941, and the property was sold by his estate to his son John Shank Habecker on 19 April. After John's death in 1972, the farm passed to his wife, its current owner, Mary Huber.

The Christian Habecker Farm retains its primary buildings, the Christian Habecker house and barn, and most of the outbuildings

The Creamery

This outbuilding served a number of functions during the history of the farm, including use as a creamery. • Ed Worteck

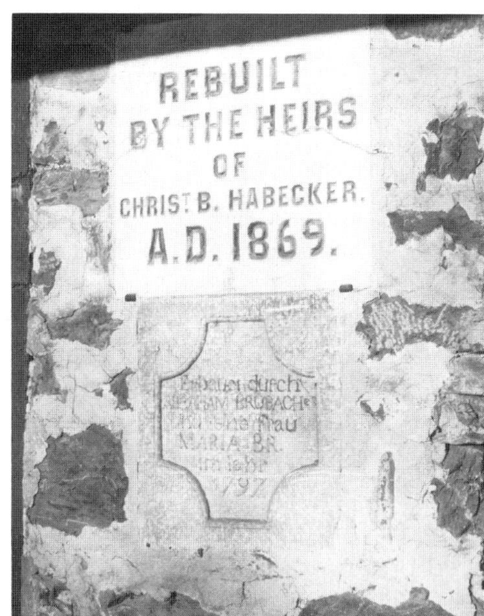

Datestones - The Barn

The datestones document the current barn and the structure it replaced. • Ed Worteck

historically associated with the Habecker family's farming operations. Domestic outbuildings, include the pigsty, a garage, a summer kitchen, and a springhouse. Two tobacco barns, a milk house, and a creamery represent the agricultural outbuildings. The site also includes a brick tenant house and its summer kitchen.

Located immediately northwest of the intersection of U. S. Route 30 and Rohrerstown Road, the Christian Habecker Farm is surrounded by residential subdivisions and retail and office facilities. In light of the changes to its surroundings, it has become increasingly difficult for the family to sustain the property as a working farm. The potential for intensive residential or commercial development on the site in the near future is significant.

The Farmhouse

A typical five-bay Pennsylvania style dwelling built by the Habecker family in 1876. *• David B. Schneider*

Farm View

This view toward the rear of the farmhouse shows its rear "tee" wing and porch. The adjacent frame pigsty is in the foreground. *• Ed Worteck*

Bausman Farmstead
Lancaster Township

The Bausman Farmstead, established in 1755, illustrates the typical evolutionary nature of Lancaster County farms and retains buildings dating from 1775 though 1879. The buildings themselves retain a high degree of architectural integrity, and each illustrates a relatively rare example of a period, style, or type of historic architecture in Lancaster.

The stone still house is the oldest of the four buildings on this farmstead. Dating from 1775, the still house was used for distilling operations through the first decades of the nineteenth century by Andreas Bausman. In 1755, Bausman moved to America from Germany and later settled on this property, which then consisted of 317 acres on either side of the present-day Millersville Pike. He had a prosperous farming and distilling business and sold his rum as far away as Pittsburgh. Although in poor condition today, the still house provides a relatively rare example of a late eighteenth-century farming related commercial structure.

Upon the death of Andreas Bausman in 1814, the property was inherited by his nephew John Bausman. John added to the property and continued to farm the land. According to its datestone, the brick dwelling on the site was built for "Jacob and Maryann Peeters" in 1836. The Peeters' relationship to Bausman and his wife Elizabeth Peters is not firmly documented.

John Bausman left this property to his son Philip in his will of 1861. Philip married Mary S. Shenk of Lancaster Township in 1859. They raised a family, continued to farm, and in 1869, according to its datestone, constructed the large brick end barn. In 1879, Philip erected a brick mansion north of the Millersville Turnpike. This elegant structure reflects the Bausman family's prosperity.

Philip's son David acquired both the mansion and the earlier structures on the farm south of the road in 1902. David Bausman was a local inventor who utilized the attic of the mansion as his workshop. Among the inventions for which he received patents are the Bausman Swing and the Bausman Wind Engine. David's daughter Edna Bausman Verdier inherited the property in 1902 and was responsible for dividing it into two separate tracts, one on either side of Millersville Pike.

The farmstead is surrounded by dense suburban development and does not retain any tillable land. The mansion house, farmhouse, and barn, even though separated by the road, are visually related and appear as a unified historic complex amid modern development.

Still House

A rare stone still house remains on the farm. • Ed Worteck

Farmhouse

The Federal style one story dwelling. • Ed Worteck

Mansion

The Bausman Mansion is a Late Victorian style dwelling possessing well detailed exterior elements such as the dormers, cornice, and porch. The interior retains its elaborate Victorian character. • *Ed Worteck*

Barn

The barn is a very large brick building with outsheds at its bank side and decorative pierced brickwork at its gable ends. It retains its decorated roof ventilators and gable vents. • *Ed Worteck*

Shreiner Farm
Manheim Township

The Shreiner Farm is an early nineteenth-century agricultural complex located on the north side of Oregon Pike, southwest of the village of Oregon in Manheim Township. The area surrounding the farm is in transition between farming and intensive suburban development. Contributing to the site are seven buildings: a stone Pennsylvania style farmhouse, a stone Pennsylvania bank barn, a frame tobacco barn, a frame and stone summer kitchen, and three frame sheds.

The farm buildings are sited on a working farm that has been owned by a single family since an original land grant from William Penn to Hans Adam Schryner in 1743. It has evolved to meet the changing needs of agricultural business, while still retaining its rural feel and fine stock of period buildings.

Hans Adam Schryner was born in Germany in 1686, and came to America in 1738. Schryner had four children, and his son Philip received the homestead upon Hans Adam's death. Philip and his wife Catherine had six children, and their son Michael succeeded to the property.

It is believed that Michael constructed the present-day farmhouse about 1830. He served in the Revolution and was a supervisor of Manheim Township. His wife Elizabeth bore four children. Henry, the only surviving child, was born in 1804, and became a school teacher. He married Mary Bard, and they had four children. Henry and Mary farmed the 106 acre property from 1827 to 1859. The farm remains in the Shreiner family.

The Shreiner Farmhouse is a representative example of an early nineteenth-century four-bay version of a Pennsylvania style farmhouse. The barn, constructed in 1828, is an excellent early nineteenth-century Pennsylvania barn. The site also includes the following contributing buildings: a frame tobacco barn, a frame and stone summer kitchen, and three frame sheds.

The Farmstead

The farmstead at the Shreiner Farm, with its three principal stone buildings, the farmhouse, barn, and kitchen, is one of Lancaster County's most picturesque farm complexes. It has received both Centennial and Bicentennial Farm designations and continues to be a working farm.
• David B. Schneider

Farmhouse

The farmhouse is typical of a four-bay Pennsylvania style farmhouse.

• David B. Schneider

Barn

The barn is a typical stone Pennsylvania barn with an 1828 datestone.

• Ed Worteck

Kitchen

The kitchen building is frame with a stone end corresponding to the interior cooking fireplace. *• David B. Schneider*

Windom Mill Farm

Manor Township

The Windom Mill Farm complex is a stately grouping of limestone buildings, including a main dwelling house, a stone end Pennsylvania bank barn, a mill, a tavern (now a dwelling), a miller's dwelling house, two tobacco sheds, a corn barn, a garage, a milk house, a pigpen, a former carriage house, and the ruinous remains of a family graveyard. The farm is a working farm and the site is located within a pristine agricultural setting.

The buildings on Windom Mill Farm illustrate a wide array of activities, including farming, commerce, and manufacturing, all important to Lancaster County agriculture from the settlement period to the present. The farm's history can be traced back to 1769, when the land on which it stands was patented to Samuel Herr. A grist and saw mill was operating as early as 1779, and a dwelling was most likely built shortly thereafter. In 1810, Herr's son-in-law, Jacob Kauffman, had the present stone flour and feed mill constructed. At about the same time, he built the tavern as a Conestoga wagon stop along Blue Rock Road, a busy commercial route in the early 1800s which led to the Susquehanna River. It seems likely the miller's house was constructed around the same time.

By 1899, a post office was operating out of the mill. The mill was also used as a creamery, and a cider mill began operating in the 1890s. The farm remained in the Landis family until 1948, when Nelson and Sylvia Landis sold approximately 106 acres to Grant and Fannie Noll. After the death of her husband, Mrs. Noll wanted to sell the farm to people she knew. So her acquaintances, Paul and Edith Rohrer, added the complex to their other farms in the area. Paul's sons leased then bought the complex from their father. P. Kenneth Rohrer, one of the sons, resides there with his family.

Both the farmhouse and the tenant house are Pennsylvania style dwellings. The miller's house is representative of the rural one-and-one-half story dwellings constructed in Lancaster County from the 1700s through the early twentieth century.

The stone barn, constructed in 1800, is a particularly fine Pennsylvania barn and the large stone mill remains architecturally intact. The site also retains two tobacco barns and a corn barn, as well as several modern buildings. Windom Mill Farm has adapted to the demands of the late twentieth-century agricultural business, yet its exceptional historic character remains and is well appreciated by its owners.

Farm View

The tavern is shown at center with the mill to the left and the farmhouse at right. • *Ed Worteck*

Farmhouse

The stone five-bay Pennsylvania style farmhouse is also representative of the Federal period. • *Ed Worteck*

Barn

The stone Pennsylvania barn on the Windom Mill Farm is exceptionally well detailed. Note the narrow ventilating slits in the gable endwall and the stone wall enclosing the barnyard. • *Ed Worteck*

Corn Barn and Tobacco Barn

Among the extensive collection of farming outbuildings are the corn barn (left) and one of two tobacco barns (right). • *Ed Worteck*

Christian & Emma Herr Farm
West Lampeter Township

The Christian and Emma Herr Farm is a working farm located in an area of intensive farming. The farm buildings, including a farmhouse, stone end barn, tenant house, summer kitchen, tobacco barn, and a shed are linearly situated along an east-west farm lane that cuts approximately through the northernmost quarter of the property.

The farm is representative of a typical Lancaster agricultural complex with architectural elements dating from the 1700s through the present. Like many Lancaster County farms, it has remained in branches of one family since the time it was established.

The oldest building on the farm is the 1761 stone end Pennsylvania bank barn. Its datestone reads, "H/1761." The tenant house, the oldest of the two brick dwellings, was built in 1864. The larger brick farmhouse was built three years later by Christian and Emma Herr and is more elaborate than is the 1864 house.

Daniel Herr, a descendant of Christian, willed the property to his daughter, Lizzie A. Herr, who subsequently passed the "Homestead farm of 100 acres in West Lampeter Township" to her son Harry B. Herr. Harry B. bequeathed the farm to his son Harry G. Herr in 1952. Harry G. Herr's son, Donald, now owns it.

Harry B. and Harry G. Herr farmed tobacco, raised steers and chickens, and kept a small dairy herd. Today, Donald farms tobacco and has a Holstein herd for dairy production. Twenty-two acres are pasture, fifteen acres are used for tobacco, and the rest of the land is used to grow crops for livestock feed.

The two dwelling houses are representative mid nineteenth-century Pennsylvania style dwellings. The barn, although altered over time, is consistent with the evolutionary nature of Lancaster County farm buildings that have been adapted to meet changing technologies and market conditions.

The Farmhouse

The farmhouse was constructed in 1867 and is representative of a five-bay Pennsylvania style farmhouse with Victorian details. Note the recessed side wing with its two-tier porch. • Heidi M. Pawlowski

The Barn

As is typical of many Lancaster County barns, the 1761 stone bank barn has been altered over time to adapt it to changing technologies and market conditions. • *Heidi M. Pawlowski*

The Tobacco Barn

• *Heidi M. Pawlowski*

Eastern Mill Creek Valley

Earl & East Earl Townships

The Eastern Mill Creek Valley is representative of the rural historic landscape that defines much of Lancaster County. Located in Earl and East Earl Townships within Lancaster County's fertile limestone plain, the valley is generally bounded to the north by the Borough of New Holland, the New Holland Pike, and the Conrail railroad tracks; to the east and south by the Welsh Mountains; and to the west by Peters Road (the Earl Township boundary). The valley contains more than 4,500 acres of prime agricultural land under active cultivation.

The National Park Service defines a "rural historic landscape" as:

A geographic area that has historically been used by people...and that possesses a significant concentration, linkage, or continuity of areas of land use, vegetation, buildings and structures, roads and waterways, and natural features.

The Eastern Mill Creek Valley fits this definition very well. Use and landscape features within the valley reflect a continuous occupation and use as a farming region for more than 250 years. The Lancaster County Historic Sites Survey has documented more than one hundred farms within the valley, many of which are owned or operated by Amish or conservative Mennonite farmers.

The farms of the Eastern Mill Creek Valley are architecturally significant, representing an excellent concentration of farmsteads dating from all periods of the county's history and reflecting the evolutionary nature of Lancaster County farms. While the integrity of each site varies, the pattern of these farms and the limited number of intrusive elements result in an area that illustrates the broad pattern of historic Lancaster County farming activities.

The Eastern Mill Creek Valley

The Eastern Mill Creek Valley is representative of much of rural Lancaster County. The pattern of its farms and circulation network has remained relatively constant for more than 250 years.

• Ed Worteck

A Farmstead in the Eastern Mill Creek Valley

While many farms within the valley are often not historically significant unto themselves, the broad collection of historic farms and the landscape, people, and traditions linking them are particularly representative of Lancaster County's nationally recognized character.

• Ed Worteck

In recent years, the Eastern Mill Creek Valley has been subjected to increasing commercial and residential development pressure. A proposed "Highway 23 Relief Route," a four-lane limited access highway that would terminate at the western end of the valley, is under consideration. Responding to extensive public concern, developers of a major retirement village proposed for the center of the valley decided to move the facility to an alternate site adjacent to New Holland Borough.

Large scale non-farm development depends largely on the availability of public sewer and water service. In 1994, an Act 537 Sewer Facilities Plan was adopted by Earl Township, designating much of the valley as an "Agricultural Preserve Non-Service Area." The purpose of the designation is to restrict access to public sewer service within the valley and thereby to control future non-farm growth. How effective these controls will be in light of other development pressures on the valley remains to be seen.

The Eastern Mill Creek Valley is representative both of the historic character of much of Lancaster County's agricultural land and of the challenges farming is facing in a rapidly growing community.

Scene in the Eastern Mill Creek Valley

(Above) Among the historic characteristics of the Eastern Mill Creek Valley are its transportation networks and the patterns in which these have been influenced by the natural environment and the activities of the people who lived here.

• Ed Worteck

A Barn in the Eastern Mill Creek Valley

(Left) The Eastern Mill Creek Valley has many farms operated by Amish and Mennonite farmers. Their farming practices and cultural patterns are a significant component of the valley's historic character and significance. *• Ed Worteck*

5: Historic Preservation

Few terms seem to conjure as many misconceptions as does "historic preservation." In fact, a great many Lancastrians are historic preservationists without even knowing it. Why? Because they own historic properties, appreciate their history, and maintain them with much the same spirit of good stewardship that has characterized Lancaster County farmers' respect for the land.

Historic preservation rarely means maintaining a property as a museum. Nor does it fail to take into account economics, changing family needs, or the availability of new technologies. Quite simply, historic preservation strives to preserve and enhance the qualities of our community that define its historic character, appearance, sense of place, and cultural traditions. Most Lancastrians share in the desire to preserve Lancaster County's beauty and quality of life. Historic preservation is, and should be, an important tool for achieving this.

It is important to remember that each and every historic property is unique in terms of its character and the influences placed upon it by its location, the needs of its owners, and its overall significance to the community. While exacting historical accuracy and conservation practices are demanded in a property like President James Buchanan's Wheatland, most historic buildings must be altered to some degree if they are to continue to be functional. It is the extent to which such changes damage the overall historic character of a property that defines whether it is sensitively preserved.

Hibschman Farmhouse

The vast majority of all historic preservation work is done by private property owners who respect and appreciate the historic character of their properties. The owners of the Hibschman Farm exemplify this. *• John P. Herr*

Restoration, Rehabilitation, Preservation

These three terms are often used interchangeably, but they are really quite different in meaning. *Restoration* is the process by which the appearance and detailing of a building are returned to exactly that of a specific period of time. In other words, the building is "put back like it was." *Rehabilitation* indicates that the condition of the building is being improved and typically adapted for a change in use or lifestyle. *Preservation* is simply the continued maintenance of the historic character and fabric of a building.

The vast majority of all historic preservation activity is defined by the latter two terms. Total restoration is generally limited to historic building museums. However, individual rehabilitation and preservation projects often involve the restoration of certain architectural elements.

It is important to understand and recognize that most old and historic buildings are owned by people who have limited financial resources and varying abilities to do work themselves. Modern conveniences like indoor plumbing, adequate kitchens, and closets are a necessity if buildings are to continue to be functionally and economically viable into the next century. The key to sensitively preserving an old or historic building is to accommodate those needs with minimal change to the building's historic character. The Historic Preservation Trust of Lancaster County can provide detailed information on ways to do this, products to use, and people who can do the work.

Christian Stauffer Farmhouse (circa 1769)

The present owners of the Christian Stauffer Farmhouse have insured the preservation of the house through the donation of a conservation easement. • John P. Herr

Historic Preservation Organizations

Several groups share responsibility for historic preservation in Lancaster County. Each functions independently of the other and, although all share common interests, the needs, concerns, and missions of each differ.

At the local level, the Historic Preservation Trust of Lancaster County, is the only countywide nonprofit membership organization actively promoting the preservation of Lancaster County's historic buildings and settings. The trust provides a variety of programs, including preservation services and technical advice, to the community. Since 1978, the trust has been compiling the Lancaster County Comprehensive Historic Sites Survey, nominating properties to the National Register of Historic Places, and assisting property owners by providing information and technical support. There are also several organizations that effectively promote historic preservation at the municipal level.

At the state level, the Pennsylvania Historical and Museum Commission, Bureau for Historic Preservation (BHP), is Pennsylvania's official State Historic Preservation Office. The BHP is responsible for implementing a variety of environmental review processes required for projects involving federal and certain state funding or licensing related to historic properties. In addition, the BHP administers the National Register of Historic Places in Pennsylvania, reviews applications for federal tax incentives that are available for the rehabilitation of historic income-producing properties, administers a variety of grant programs, and conducts other programs and activities. Preservation Pennsylvania is a nonprofit organization that promotes historic preservation statewide.

The National Trust for Historic Preservation is a nationwide nonprofit advocacy group that was chartered by Congress in 1949. The National Park Service is a federal agency involved in the National Register, tax incentives, technical assistance, and other programs.

Abraham Landis Farmhouse (circa 1763)

High Industries rehabilitated this significant historic building, successfully converting it from a dwelling into offices while preserving its character defining elements. • *John P. Herr*

Designating Historic Buildings

Professional criteria have been developed for use in evaluating and designating historic buildings. These criteria apply to both the National Register of Historic Places and to the Lancaster County Historic Sites Register.

The National Register of Historic Places is this nation's official listing of buildings, structures, sites, districts, and objects that possess significance in American history, archaeology, engineering, and/or culture. Designation in the register does not restrict a property owner in any way and provides limited protection from demolition or alteration by federal and state projects.

The Lancaster County Historic Sites Register is a program administered by the Historic Preservation Trust of Lancaster County. While based largely on the National Register, the Lancaster County register has been adapted to provide increased recognition for sites of predominantly local significance.

Both designations recognize buildings, structures, sites, districts, and objects that possess significance in national, state, or local history, architecture, and culture. In addition, sites must continue to have integrity, or the ability to convey the overall essence of their historic location, design, setting, materials, workmanship, feeling and associations.

Incentives and Preservation Tools

A federal investment tax credit is available for the rehabilitation of income producing buildings listed on the National Register of Historic Places. In the case of a farm, this would generally include any buildings used in agricultural activities that produce income for the farmer. The credit equals twenty percent of the amount actually spent rehabilitating a building and is applied directly to federal income tax owed. Rehabilitation work must conform to *The Secretary of the Interior's Standards for Rehabilitation*, a set of practical guidelines established by the National Park Service to insure such projects are sensitive to the historic character of the buildings involved.

Conservation easements are a second important preservation tool that can provide tax and estate planning benefits. A conservation easement is basically a voluntary, legally binding agreement between a property owner and a qualified conservation organization where the the right to demolish or substantially alter a historic building or buildings is restricted.

Anyone interested in utilizing any of the tax incentives may contact the Historic Preservation Trust and should consult a qualified tax advisor.

Regulation

"If my building is declared historic, I won't be able to change a thing." This is a fear some people have expressed when they hear the words historic preservation. It reflects their concerns about the potential designation of their properties as historic sites and of municipal attempts to preserve historic buildings. When such objections do arise, they are often due to misinformation.

Historic preservation is a legitimate community interest that is specifically mentioned in the Pennsylvania State Constitution. Historic preservation regulations are an accepted planning tool for local governments. A recent state Supreme Court ruling has added further legitimacy to this concept and many Lancaster County municipalities already have preservation provisions in their planning and zoning ordinances.

Historic preservation provisions are often the only effective way to insure that the historic character and resources of a municipality are not destroyed. Such ordinances can provide substantial incentives to promote private preservation initiatives, rather than simply restricting a property owner's ability to make alterations.

The degree to which a preservation ordinance controls changes to, or demolition of, historic buildings varies. It is important to recognize that, other than the review of federal, (and some state) funded or licensed activities, such regulations are adopted at the discretion of the municipality and its elected officials. Therefore, historic preservation ordinances are typically tailored to the specific needs of the municipality and its residents.

For additional Information

Historic Preservation Trust of Lancaster County
123 N. Prince Street, Lancaster, PA 17603
Phone: (717) 291-5861

When Preservation Fails

Lancaster County loses more and more elements of its nationally recognized historic agricultural character each year. Gradually, the traditional qualities of our sense of place are being destroyed.

• Historic Preservation Trust

Notes

1: Farming in Lancaster County

[1] A precise origin for this nickname has never been fully documented. Pennsylvania was often referred to in early accounts as the "Granary to the Colonies." Reference to the use of the title "Garden Spot of America" for Lancaster County can be dated to at least 1853, when Eli Bowen's *The Pictorial Sketch-Book of Pennsylvania* states: "An intelligent Englishman called this county the 'Garden of America....' [it] is, without a doubt, the garden of this glorious Union, and there are few spots in this wide, wide world, which could present a nobler scene to the eye than is here afforded."; Eli Bowen,*The Pictorial Sketch-Book of Pennsylvania* (Philadelphia: William Bromwell, 1853), p. 31.

[2] The term "Pennsylvania Dutch" appears to have originated in the nineteenth century as a mispronunciation of "Pennsylvania Deutsch" which refers to the county's large Pennsylvania German community. The earliest migrations of Europeans from what is now Germany began to arrive in America in the early eighteenth century and developed a Germanic culture that actually predated the rise of German nationalism. Pennsylvania Germans therefore developed many of their own traditions and dialects. As Germany began to take on the aspects of a unified nation in the nineteenth century, later immigrants from the region arrived with a more unified cultural tradition and formed the group we refer to as German-Americans.

[3] H. M. J. Klein, Ph.D., ed., *Lancaster County Pennsylvania: A History* (New York: Lewis Historical Publishing Co., Inc., 1924), p. 4.

[4] John Fraser Hart, The Land That Feeds Us (New York: W. W. Norton & Company, 1992), p. 20.

[5] Franklin Ellis and Samuel Evans, *History of Lancaster County, Pennsylvania*. 2 Volumes (Philadelphia: Everts and Peck, 1883), Volume I, pp. 6-18.

[6] Stevenson Whitcomb Fletcher, *Pennsylvania Agriculture and Country Life 1640- 1840* (Harrisburg, PA: Pennsylvania Historical and Museum Commission, 1971) , p. 4

[7] The Anabaptist movement developed in the sixteenth century, a time of significant social and religious transformation. John A. Hostetler provides a succinct description of the roots of the Mennonite, and subsequently the Amish, sect in *Amish Society* (Baltimore: The Johns Hopkins University Press, 1980), pp. 25-27; see also Donald B. Kraybill's *The Riddle of Amish Culture* (Baltimore: The Johns Hopkins University Press, 1989), pp. 1-8.

[8] Ellis and Evans, pp. 9, 324-337;

[9] Ellis and Evans, pp. 9; J. Russell Cross, *Historic Ramblin's through Berkeley* (Columbia, S.C.: The R. L. Bryan Co., 1985), pp. 106-117.

[10] H. M. J. Klein, p. 16.

[11] Ellis and Evans, pp. 9-IO.

[12] Kraybill, p. 8.

[13] Ellis and Evans, p. 345.

[14] Ivan Glick, "The Story of Barns," *Ford New Holland News*. June I991, pp. 6-22.

[15] Fletcher, pp. 127-139.

[16] Fletcher, p. 133; "Valley of the Pequea" probably refers to the area of present-day Upper Leacock or Earl Townships.

[17] Ellis and Evans, p. 348-351.

[18] Fletcher, p. 123.

[19] Lemon, p. 79.

[20] Fletcher, p. 57.

[21] Berks County Conservancy, "Farms in Berks County, PA" multiple property nomination to the National Register of Historic Places, 1992, p. E-6.

[22] Fletcher, p. 199-200.

[23] Fletcher, p. 117.

[24] Ellis and Evans, pp. 301-305; Historic Resource Survey files for Hopewell Forge, Speedwell Forge, Windsor Forge, and Martic Forge, Historic Preservation Trust of Lancaster County.

[25] Arthur C. Lord, "Architectural Characteristics of Houses: Lancaster County, pp. 132-151.

[26] H. M. J. Klein, p. 660; U. S. Census statistics.

[27] Hart, p. 25.

[29] Ellis and Evans, p. 350.

[30] Worner, pp. 99-100

[31] Hart, pp. 21-22.

[32] Fletcher, p. 181.

[33] H. M. J. Klein, p. 667.

[34] Fletcher, p. 187-188.

[35] H. M. J. Klein, p. 667.

[36] Frederic Klein, p. 18.

[37] Willis N. Baer, Ph.D.,*The Economic Development of the Cigar Industry in the United States* (Lancaster, PA: n.p., 1933); William Frear, Ph.D., "The Cigar Tobacco Industry in Pennsylvania," Pennsylvania Dept. of Agriculture, General Bulletin No. 371 (1922), in Bulletins, 5 (Nos. 364-371; PA 12), State Library of Pennsylvania, Government Publications Division; Horace Richards Barnes, "Early History of Tobacco," in *Papers of the Lancaster County Historical Society*, 45 (1942): 1-24, 6; Fletcher, 165-6.

[38] Fletcher, 166; Ellis and Evans, 355; Baer, 38.

[39] Baer, 37-8; Good, 194-7; National Register nominations for the North Prince Street Historic District, the North Charlotte Street Historic District, and the Nissley-Stauffer Tobacco Warehouses, all Lancaster City. Figures for Lancaster County tobacco yields vary somewhat depending on the source, but in general appear reliable; figures for 1936 are from Barnes, 7, citing the *Yearbook of Agriculture, 1935*.

[40] Barnes, 14, citing *The Lancaster Farmer*, XV, 2 (Feb. 1883), 23; Ellis and Evans, I:355.

[41] Good, 196-7.

[42] Geoffrey H. Ranck of Domestic Tobacco Co., Lancaster, PA, provided much information on the growing, curing, and packing of leaf tobacco, as well as the fermenting and repacking at the warehouse (below); telephone conversations with Gloria 0. Becker, Ph.D., 1989. Also, Joe Fanelli, Packing Manager, Lancaster Leaf Co., provided informa-

tion about the leaf packing process and in particular, about the buildings at 850 N. Water Street; further useful information on drying and curing of tobacco is in Frear, 28-62.

[43] H. M. J. Klein, p. 662.

[44] Fletcher, p. 257.

[45] 1860 and 1870 Census statistics.

[46] Fletcher, p. 165.

[47] Fletcher, p. 167.

[48] Fletcher, p. 183.

[49] Frederic Klein, p. 19.

[50] Fletcher, p. 54.

[51] Gideon L. Fisher, *Farm Life and its Changes* (Gordonville, PA: Pequea Publishers, 1978), p. 52

[52] Lancaster County Farmers Directory for 1909; Frederic Klein noted the comparison in size between Lancaster at 4,202 square miles and Los Angeles County at 7,326 square miles (Frederic Klein, p. 87).

[53] 1900 Census.

[54] Frederic Shriver Klein, *Lancaster County 1841-1941* (Lancaster, PA: Intelligencer Printing, 1941), pp. 154-155.

[55] Fisher, p. 16.

[56] Frederic Klein, p. 186.

[57] See Note 39.

[58] Fletcher, pp. 119-126; Interview, Steven Miller, Pennsylvania Farm Museum at Landis Valley, Lancaster PA, 25 March 1994.

[59] H. M. J. Klein, p. 667.

[60] Fletcher, pp. 207-210.

[61] H. M. J. Klein, p. 678; Fletcher, p. 234.

[62] Fletcher, pp. 239-255.

[63] Scott, pp. 74, 78-80.

[64] In 1994, only Central Market remains in active use as such a market.

[65] Gary Hovinen, "Lancaster's Streetcar Suburbs, 1890-1920," *Journal of the Lancaster County Historical Society*. Vol. 82 (1978), pp. 50-51; Frederic Klein, p. 118-119.

[66] Frederic Klein, p. 196-197.

[67] Historic Preservation Trust of Lancaster County, Historic Sites Survey Files, "Conestoga House," research by John J. Snyder, Jr., 1981.

[68] Lancaster County Planning Commission. *Lancaster County Comprehensive Plan: Growth Management Plan* (Lancaster PA: Lancaster County Planning Commission, Draft, January 1992), p. 5.

[69] Lancaster Tomorrow Group, *Lancaster Today: A Description of Community Life* (Lancaster PA: The Lancaster Tomorrow Group, 1977) p. 26.

[70] Lancaster Tomorrow Group, p. 26.

[71] Ed Klimuska, "Lancaster County: The (Ex?) Garden Spot of America," *Lancaster (PA) New Era*, 1988, referenced in John A. Hostetler, "Toward Responsible Growth and Stewardship of Lancaster County's Landscape," *Pennsylvania Mennonite Heritage*, Vol. XII, No. 3 (July 1989), p. 3.

[72] Lancaster County Planning Commission. *Lancaster County Comprehensive Plan: Policy Plan* (Lancaster PA: Lancaster County Planning Commission, 1991), p. 3.

[73] Lancaster County Planning Commission. *Lancaster County Comprehensive Plan: Growth Management Plan* (Lancaster PA: Lancaster County Planning Commission, Draft, April 1993), p. 3.

[74] Frederick Klein, p. 113.

[75] Lancaster County Planning Commission, *Directions: A Comprehensive Plan for Lancaster County, Pennsylvania* (Lancaster, PA: Lancaster County Planning Commission, 1975), pp. 1-5.

2: The Agricultural Landscape

[1] William Frederick Worner *Old Lancaster: Tales and Traditions* (Lancaster, PA: William Frederick Worner, 1927), p. 37.

[2] Ibid.

[3] Ibid, p. 57.

[4] Ibid, p. 107-108.

[5] Ibid, p. 114.

[6] Ibid, p. 150-151.

[7] Ibid, p. 163.

[8] Eli Bowen, *The Pictorial Sketch-Book of Pennsylvania* (Philadelphia: William Bromwell, 1853), pp. 31-32.

[9] Worner, p. 58, 78, 114, 164.

3: Farm Buildings in Lancaster County

[1] Arthur C. Lord, "Architectural Characteristics of Houses: Lancaster County, 1798," *Journal of the Lancaster County Historical Society*, Vol. 85, No. 4 (1981), pp. 132-151.

[2] According to research provided by Lancaster County historian and author Henry J. Kauffman, the Flurküchen plan would actually represent somewhat of a variation of the Herr House plan. Where the Küche is partitioned at its rear to provide a workroom in the Herr House, a traditional German Flurküchen plan would have a small room partitioned to the front of the Küche as an entry hall, or Flur. Often the original German precedents also have an attached stall or stable at the opposite end from the Stube and Kammer. Only one house has been documented in Lancaster County with such a plan, the Reiber-Herr-Hershey House near Creswell. This particular house is rather late for a "pure" Germanic house in this county as it dates from circa 1772 and additional research is required to determine if the existing plan is indeed original.

[3] Berks County Conservancy, "Farms in Berks County, PA" multiple property nomination to the National Register of Historic Places, n.p., 1992.

[4] Joseph W. Glass, *The Pennylvania Culture Region: A View from the Barn* (Ann Arbor, MI: UMI Research Press, 1971, 1986), pp. 111-169.

[5] Henry J. Kauffman, *Architecture of the Pennsylvania Durch Country, 1700-1900* (Lancaster, PA: By the author, 1992), p. 59-94.

[6] A "Dawdy Haus," or grandfather house, is typically defined as either an attached addition or separate structure built to house the older generation of a family when the younger generation assumes principal responsibility for the operation of the farm. In common practice, Dawdy Haus additions may also house other generations of the family as well. In many cases, successive additions are constructed and result in houses with stepped or telescoping appearance.

Bibliography

Arthur, Eric, and Dudley Witney. *The Barn.* Ontario: M.F. Feheley Arts Company Limited, 1972.

Berks County Conservancy. "Farms in Berks County, PA" multiple property nomination to the National Register of Historic Places. n.p., 1992.

Biographical Annals of Lancaster County. Spartanburg, SC: Reprint Publishers, 1989.

Brumbaugh, G. Edwin. *Colonial Architecture of the Pennsylvania Germans.* Norristown, PA: Pennsylvania German Society, 1933.

Bucher, Robert C. "Steep Roof and Red Tiles." *Pennsylvania Folklife.* Lancaster, PA: Pennsylvania Folklife Society, Summer 1961.

____. "The Swiss Bank House in Pennsylvania." *Pennsylvania Folklife.* Vol.18, 1969.

Dornbusch, Charles H., and John K. Heyl. *Pennsylvania German Barns.* Allentown, PA: The Pennsylvania German Folklore Society, 1958.

Ellis, Franklin, and Samuel Evans. *History of Lancaster County, Pennsylvania,* Volumes I & II. Philadelphia: Everts and Peck, 1883.

Ensminger, Robert F. *The Pennsylvania Barn.* Baltimore: The Johns Hopkins University Press, 1992.

Everts and Stewart. *The 1875 Historical Atlas of Lancaster County, Pennsylvania*; reprint ed. Knightstown, IN: The Bookmark, 1976.

Fegley, H. Winslow. *Farming, Always Farming.* n.p. The Pennsylvania German Society, 1986.

Fisher, Gideon L. *Farm Life and Its Changes.* Gordonville, PA: Pequea Publishers, 1978.

Fletcher, Stevenson Whitcomb. *Pennsylvania Agriculture and Country Life 1640-1840.* Harrisburg, PA: Pennsylvania Historical and Museum Commission, 1971.

_____. *Pennsylvania Agriculture and Country Life 1840-1940.* Harrisburg, PA: Pennsylvania Historical and Museum Commission, 1971.

Glass, Joseph W. *The Pennsylvania Culture Region: A View from the Barn.* Ann Arbor, MI: UMI Research Press, 1986.

Hart, John Fraser. *The Land That Feeds Us.* New York: W. W. Norton & Company, 1992.

Historic Preservation Trust of Lancaster County. *Our Present Past.* Lancaster, PA: The Historic Preservation Trust of Lancaster County, 1985.

____. *Lancaster County Architecture 1700-1850.* Lancaster, PA: The Historic Preservation Trust of Lancaster County, 1992.

Hostetler, John A. *Amish Society.* Baltimore: The Johns Hopkins University Press, 1980.

____. "Toward Responsible Growth and Stewardship of Lancaster County's Landscape." *Pennsylvania Mennonite Heritage,* July 1989.

Jackson, John Brinckerhoff. *Discovering the Vernacular Landscape.* New Haven, CT: Yale University Press, 1984.

Kauffman, Henry J. *Architecture of the Pennsylvania Durch Country, 1700-1900.* Lancaster, PA: By the author, 1992.

Klein, Frederick Shriver. *Lancaster County 1841-1941*. Lancaster, PA: The Lancaster County National Bank, 1941.

Klein, H. M. J. *Lancaster County, Pennsylvania: A History*. 4 vols. New York: Lewis Historical Publishers, 1924.

Kollmorgen, Walter M. *Culture of a Contemporary Rural Community*. Washington, D.C.: U.S. Department of Agriculture, Bureau of Agricultural Economics, 1942.

Kraybill, Donald B. *The Riddle of Amish Culture*. Baltimore: The Johns Hopkins University Press, 1989.

Lancaster County Planning Commission. *Lancaster's Heritage*. Lancaster, PA: Lancaster County Planning Commission, 1972.

Lemon, James T. *The Best Poor Man's Country: A Geographical Study of Early Southeastern Pennsylvania*. New York: W. W. Norton & Company, 1976.

Loose,John W. W. *The Heritage of Lancaster*. Woodland Hills, CA: Windsor Publications, 1978.

Long, Amos. *The Pennsylvania German Family Farm*. Breinigsville, PA: The Pennsylvania German Society, 1972.

Lord, Arthur C. "Architectural Characteristics of Houses in Lancaster County, 1798," *Journal of the Lancaster County Historical Society*, Volume 85 (1981): 132-151.

____. "Barns of Lancaster County: 1798," *Journal of the Lancaster County Historical Society*, Volume 77 (1973): 26-40.

Meining, D.W., ed. *The Interpretation of Ordinary Landscapes: Geographical Essays*. New York: Oxford University Press, 1979.

Raymond, Eleanor. *Early Domestic Architecture of Pennsylvania*. Exton, PA: Schiffer Ltd., 1930.

Richman, Irwin. *Pennsylvania's Architecture*. University Park, PA: The Pennsylvania Historical Association, 1969.

Scott, Stephen. *Amish Houses & Barns*. Intercourse, PA: Good Books, 1992.

Shoemaker, Alfred L. *The Pennsylvania Barn*. Lancaster, PA: Pennsylvania Dutch Folklore Center, Inc., 1978.

Sloane, Eric. *An Age of Barns*. n.p.: , n.d.

Stilgoe, John R. *Common Landscape of America, 1580 to 1845*. New Haven and London: Yale University Press, 1982.

Upton, Dell, and John Vlach, eds. *Common Places: Readings in American Vernacular Architecture*. Athens, GA: University of Georgia Press, 1980.

Weaver, William Woys. "'Weizenthal' and the Early Architecture of Neu-Strasburge: Swiss Plantations in the Province of Pennsylvania." Ph. D. dissertation, University of Virginia, 1969.

Worner, William Frederick. *Old Lancaster: Tales and Traditions*. Lancaster, PA: William Frederick Worner, 1927.

Epilogue

• Ed Worteck

Will We Pass Our Inheritance on to Our Children?

The physical legacy of farming in Lancaster County is disappearing. Each passing year sees the loss or substantial alteration of numerous buildings and settings. Between 1975 and 1981, more than fifteen percent of the historic sites recorded in *Lancaster's Heritage* were lost forever. In the past ten years, more than sixty-eight square miles of Lancaster County were approved for subdivision. Rising land prices and the shrinking availability of agricultural land are causing plain sect farmers to leave the county. Many of the historic farming related buildings illustrated in this book no longer stand and others will undoubtedly be lost soon.

Lancaster County's farming related buildings and settings are a tangible reminder of the people who carried on its rich traditions for almost three hundred years. Yet our generation may well be the one to squander this inheritance. Will Lancaster County's important legacy of farming architecture survive? The answer to the question, unlike so many of the buildings, is not set in stone. If we hope to save this very special place, Lancastrians must actively support both historic and agricultural preservation efforts. The Historic Preservation Trust and other community organizations need your time, talents, and contributions.

We have the power and the responsibility to pass this significant historic place and its unique quality of life on to future generations.

The following farming related buildings illustrated in this book have been lost forever:

Page 53	John Herr Farmstead, 1813	Replaced by a golf course
Page 53	Jacob Siegrist Farmhouse, 1822	Demolished, site vacant.
Page 54	William & Mary Smith Farmhouse, 1853	Replaced by a parking lot
Page 54	Daniel Kurtz Farm, 1861	Replaced by a medical facility.
Page 55	Christian O. Herr Farmhouse, 1860	Demolished, site vacant.
Page 57	Jacob Esbenshade Farmhouse, ca. 1915	Replaced by a retail store.
Page 59	Jacob Shank Barn, 1835	Demolished, site vacant.

Demolition of the Esbenshade Farmhouse

Historic buildings were typically well constructed and often prove to be adaptable for new uses. Tens of thousands of old and historic buildings are lived and worked in every day in Lancaster County. Some simply continue to serve their original purpose, as dwellings, farm buildings, stores, etc., and others have been successfully rehabilitated for new uses. Yet we often take for granted their presence. Contrary to what some people think, there are very few instances where the preservation of individual historic buildings is no longer viable, either because of severe deterioration or inability to meet present day needs.

• Lori Stahl